Rain and Ruin

The Diary of an Oxfordshire Farmer
John Simpson Calvertt
1875–1900

John Simpson Calvertt, 1899. (By kind permission of the executors of Mrs. D.E. Hall.)

Rain and Ruin

The Diary of an Oxfordshire Farmer
John Simpson Calvertt
1875–1900

Edited
by

Celia Miller

ALAN SUTTON
1983

Alan Sutton Publishing Limited
17a Brunswick Road
Gloucester

First published 1983

ISBN 0 86299 113 7 (Case)
ISBN 0 86299 114 5 (Paper)

Typesetting and origination by
Alan Sutton Publishing Limited.
Photoset Bembo 10/11
Printed in Great Britain.

Contents

Acknowledgements

The diary of John Simpson Calvertt first came to my notice in 1976 and I resolved then to bring the man, his life and times to a wider audience at some future date. Unfortunately, the person who would have been most pleased to see this book died a few months before Alan Sutton agreed to publish it. Mrs. Dorothy Esmé Hall's birth on 13 July 1895 was recorded by her grandfather in his diary and her connection with it remained unbroken after his death in 1900. First her father, J.S. Calvertt's eldest son, John Charles, and then Mrs. Hall herself were responsible for the preservation of the surviving volumes of the diary until they were deposited in the care of the County Archivist for Gloucestershire. Mrs. Hall gave her enthusiastic approval to my suggestion that I should edit her grandfather's diary for publication, since she also thought that it merited wider exposure. She died on 22 October 1982, so near to the achievement of her wish. Without her care of the diary over so many years it might never have survived – I am, therefore, greatly indebted to her. I am also grateful to Mrs. Hall's executors for their permission to continue with the project.

I must also thank D.J.H. Smith, the County Archivist, and the staff of the search room at Gloucestershire Record Office for their practical help and friendly assistance over many years. The staff of the Local History Library at Westgate Central Library, Oxford were also most helpful in the search for photographic material and I am grateful for permission to reproduce some of the photographs.

Friends and colleagues have made their contributions, both personal and professional, to the production of this book. Amongst them, David Viner merits particular thanks for his help. My thanks are also due in no small measure to my mother for her encouragement and unflagging support. Above all, I owe a very large debt of gratitude to my husband, without whose help, encouragement and fortitude in the face of Calvertt-induced domestic crises this book would certainly never have seen the light of day. A man of many parts, he is also responsible for the maps and some of the photographs.

Introduction

John Simpson Calvertt was born on 13 November 1829 at Withern, Lincolnshire, the son of John Calvertt, surgeon apothecary of Withern. His father died of tetanus in 1846 and the future of John and his younger brother, Charles, was left in the hands of three trustees. In 1847 he went as an agricultural pupil to Mr. Richard Adams of Collow Grange, Collow in Lincolnshire, where he spent two happy years. From July 1849 he lived and worked with one of his trustees, Mr. William Barker, on the Barker farm at Claythorpe before entering Tothill Manor Farm, Alford as a tenant of Lord Willoughby de Broke in April 1852.

He married Jane Waters Fowler (born 4 May 1832), who came from a Louth farming family, at Louth on 13 June 1854. Six children, three girls and three boys, were born at Tothill during the couple's twenty-one year residence there. The decision to move from Lincolnshire to Oxfordshire arose from Calvertt's desire to farm on a much larger scale than was possible on the holdings available to him in his native Lincolnshire in 1875, together with the fact that his Tothill land was liable to frequent flooding. The move from Tothill Manor Farm to a Crown estate holding: Fairspear and High Lodge Farms, Leafield, near Shipton-under-Wychwood, Oxfordshire, took place between October 1875 and April 1876. Calvertt remained a tenant of the Crown until his death on 2 May 1900, his wife having predeceased him in 1889.

Calvertt began to keep a diary as a schoolboy at Cresswell's Northgate Academy in Louth. This record consisted originally of five volumes covering the years from 14 July 1845 to 31 March 1900, but unfortunately the third volume, covering the period from 1 January 1867 to 31 December 1874 is now missing. This loss not only spoils the continuity of the story of a long and full farming life, but it also detracts from the value of what is otherwise an unusually long and detailed primary source of nineteenth century agricultural history. Fortunately, the last two volumes, which run from 1 January 1875 to 31 March 1900 are a complete record of Calvertt's life in Oxfordshire and are not affected by the loss of the previous volume.

The village of Leafield is situated on the west Oxfordshire limestone uplands between the valleys of the Evenlode and the Windrush. It grew up originally in a clearing in the great royal forest of Wychwood but the surrounding landscape was radically changed when an Act for the disafforestation of Wychwood was passed in 1857, and now only a remnant of the forest survives in Cornbury Park. The Crown was allocated some 2543 acres under the enclosure award, augmented by exchange and purchase to 2937 acres. About 1970 acres of this Crown holding was unreclaimed forest, but in an incredibly short space of time – less than two years between October 1856 and January 1858 – the land had been completely cleared of trees and laid out into seven farms which were let to tenants who had tendered for the leases. The seven farms: Ascott, High Lodge, Fairspear, Langley, Potters Hill, South Lawn, and Leafield, totalled approximately 2843 acres in extent. All the farms lay, for the most part, on the thin Cotswold-type soil known as Stonebrash, with a smaller proportion of heavier Cornbrash soil and some Oxford Clay. Stonebrash is a hungry soil: easy to work and productive when extensively fertilised, but greedy for water; in very hot weather or in drought conditions Calvertt found that temporary grass crops and turnips dried up and were quickly ruined. Adequate water supplies for stock were also a permanent problem on some of the farms. The original tenant of Fairspear and High Lodge Farms was a Mr. Buckle, but by the time Calvertt took the lease in 1875 the farms had already changed hands once. Two changes of tenancy in the space of seventeen years would seem to indicate that cultivation of the land was not without problems. High Lodge seems to have proved more difficult than Fairspear, for it was the High Lodge tenancy that Calvertt relinquished when his seventeen year lease expired in 1892, taking up instead the tenancy of the adjacent Langley Farm. The tenancy of Fairspear was transferred to his eldest son, John Charles, in 1894, and Calvertt moved to Langley Farm, where he lived until his death in 1900.

The diary left behind by Calvertt is a record of his world at every level: business and economic life, economic, social and family relationships, local government, political activities and views, national and world events and, above all, his own perception of all these things. The last quarter of the nineteenth century was a period of accelerating political, social and economic change, when many of the accepted values and institutions of the preceding decades of the century were being challenged by powerful internal and external influences. The nature of these influences is reflected in Calvertt's diary entries, of which a selection are reproduced in the pages of this book. The abbreviated and colloquial style of the author, and his often eccentric punctuation and spelling may seem a little strange initially but these features are so much part of the man that they have been left unaltered, except where they actually interfere with the reader's understanding of the text. The historical background to the diary is varied and some explanation of it might prove useful before reading the contents.

Calvertt's decision to move to Oxfordshire came at a point in time when British agriculture was embarking upon a period of prolonged depression. The years stretching from the 1840s through to the mid-1870s, known generally as the 'golden age' of farming, were years of relative prosperity in agriculture: high production achieved through intensive cultivation and good management, aided by rising livestock and dairy produce prices, and the application of agricultural science and technology. These increased production levels were stimulated and sustained by the expansion of urban markets, gradually rising real incomes, and the growth of the railway system. The last quarter of the nineteenth century presents an apparently sudden contrast and was quickly christened the period of the 'great depression' in British agriculture. Its main characteristics were rising imports of agricultural produce, falling grain prices, a run of generally bad weather in the 1870s and 1880s, disease amongst farm stock, and financial distress amongst landowners and farmers.

Some features of the depression had their origins in the preceding period of prosperity. The balance of consumer demand, for instance, had been shifting away from cereals and towards livestock and dairy produce for much of the century. Imports of agricultural produce had been steadily increasing ever since the removal in 1846 of the restraints imposed by the Corn Laws, although import levels rose still further during the last three decades of the century once the difficulties involved in bulk transport over long distances were gradually overcome. Disease in the form of the rot (liver fluke) had been endemic for years, and other epidemic diseases such as foot-and-mouth, rinderpest (cattle plague) and pleuro-pneumonia had arrived in Britain during the 1860s with increased imports of live animals. But other features of the depression were new: corn prices, although they had periodically fallen to low levels after 1846, fell permanently in the late 1870s. Allied to the shift in consumer demand away from the cereal sector this development was obviously a recipe for disaster in predominantly grain-growing areas, although livestock farmers could use cheap grain as animal feed. Raw wool prices fell drastically during the 1880s, thus nullifying the benefit of any expansion in sheep farming because the more profitable meat end of the business merely subsidised the losses made on wool. The sharp fall in the prices of meat and dairy produce which occurred in the early 1890s came as a direct result of improvements in shipping techniques, particularly the development of refrigeration during the 1880s, and a consequently increased volume of imports. This new wave of falling prices, following as it did on the earlier fall in grain and wool prices, constituted a very real threat to the profits of both arable and grassland farmers. Only liquid milk production remained immune from foreign competition and for this reason it is not surprising that much of the technical innovation of the late 1890s and the present century has been centred on the dairy industry.

The agriculture of nineteenth century England can be divided into two very broad categories: the southern and eastern counties where the classical mixed farming system based on corn, with sheep as a subsidiary crop, was predominant; and the northern and western grassland counties where the emphasis was on dairying and livestock. Generally speaking, the depression affected the southern and eastern counties more than the northern and western counties. But this geographical divide was complicated by regional sub-divisions and no area of the country was completely immune from all aspects of the depression, for livestock and dairy areas were affected by the frequently chaotic weather conditions and by disease amongst stock.

The weather was an important factor during the depression. The run of erratic and extreme conditions began in 1875 with a cold wet summer and persisted with only short interruptions through to the mid-1890s with cold wet springs and summers, floods alternating periodically with droughts, and with some extreme winter conditions, notably the late snow of spring 1879, the blizzard of 1891, and the bitter cold of the winter of 1894–5. The same period of years was marked by heavy losses of stock from intermittent outbreaks of liver fluke, induced by wet conditions, and of pleuro-pneumonia and foot-and-mouth disease. It has been the practice of some historians to minimise the effects of the great depression in agriculture between 1875 and 1895 by emphasising its regional nature, but there can be little doubt that farmers everywhere, on every type of land, and using every type of management system felt the effect of falling prices, extreme weather conditions and disease amongst stock. It was only the *extent* to which farmers suffered from the effects which varied regionally.

This is the background against which the last two volumes of Calvertt's diary open. Calvertt followed the classical mixed farming system: flexible rotations of corn and fodder crops with temporary grasses, and large flocks of sheep (Oxford Down and Cotswold) fed on the fodder crops in the field, thus fertilising the ground at the same time. In addition, he raised and fattened cattle in yards and pastures, thus providing more valuable manurial residues. The various problems posed by conditions during the depression were therefore encountered and recorded by Calvertt in his diary.

But the physical and economic characteristics and extent of the agricultural depression were only part of what was happening to farmers at that time. To them, it was bad enough that the forces of the weather and foreign imports were apparently conspiring to erode their economic well-being, but it must also have seemed that their political and social status was being assaulted by the effects of government policy and electoral and administrative reform. The landed interest, that powerful alliance of landowners and tenant farmers, had for many years prior to the 1830s held the government of the country at national and local level in the palm of its hand. This seemingly invincible position had been persistently eroded by successive social and political reforms, beginning with the Reform Act of 1832 and the New Poor Law of 1834, gradually transferring power from the hands of the landed interest to representatives elected by an ever-widening electorate. This process was accelerated by the Reform and Redistribution of Seats Acts of 1884–5 which finally extended the franchise to rate-paying householders in the counties, a privilege extended to borough ratepayers in 1867. At local level the landed interest had governed the counties through non-elected Justices of the Peace in petty and quarter sessions, and the parishes through non-elected parish vestries, plus *ad hoc* bodies such as education and sanitary boards set up piecemeal to perform specific functions. This confused and far from democratic situation was partly remedied by the Local Government Act of 1888 which transferred the administration of counties, including the levying of rates, maintenance of roads, bridges, lunatic asylums and poor relief to county councils elected by ratepayers.

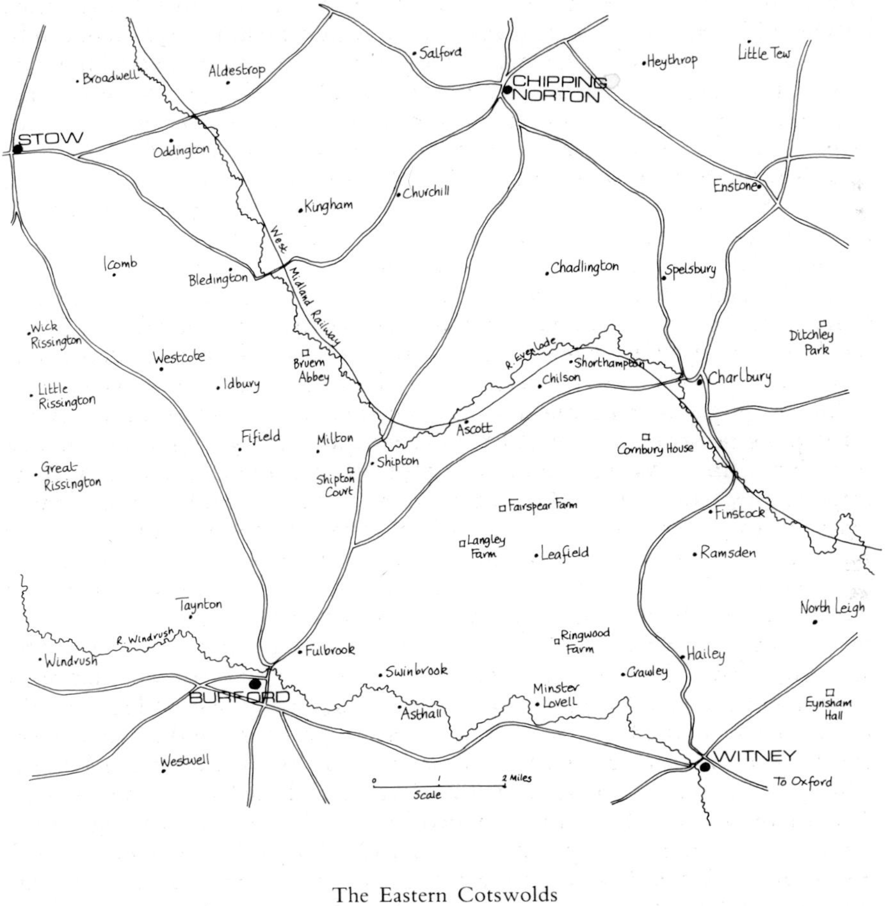

The Eastern Cotswolds

The Local Government Act of 1894 went several steps further by giving the vote in local elections (previously based on a property qualification) to all county and parliamentary electors, and then further delegating local government by creating rural and urban district councils and transferring the civil functions of parish vestries to specially-created parish councils. This democratisation of the franchise at national and local level and the transfer of power at county, town, district and parish levels from self-perpetuating to elected bodies clearly militated against the landed interest. The declining importance of agriculture in an increasingly industrial economy, taken together with waning social and political influence and evident economic distress, had inevitable consequences. Agriculture became, and has remained a secondary consideration at government level. Not until the demands made by war, first in 1914–18 and then, more significantly, in 1939–45 was the creeping decline in British agriculture arrested.

The social repercussions of the economic situation in agriculture and the progress of the transfer of power away from the landed interest during the last quarter of the nineteenth century were frequently drastic and they were recorded by Calvertt with a mixture of relish and despondency. Farmers in Calvertt's part of west Oxfordshire seem to have been particularly hard-hit by the great depression and the incidence of farming bankruptcies (or 'failures') in the diary is frequent, but whether all these were actual bankruptcies, or simply decisions to give up farming after heavy losses is a debatable point. The progress of the depression can also be traced in the continual remissions or returns of rent paid made to Calvertt by the Crown during the late 1870s and 1880s. The rent of Fairspear Farm was renegotiated in 1892 when the original seventeen year lease expired. The Crown asked for a rent of £600 for the 660-acre farm but Calvertt eventually negotiated a rent of £450 per year – good tenants with a sound financial background and a proven record were obviously hard to find. But the Crown was an extremely generous landlord and Calvertt seems to have received a great deal more help through the difficult depression years than his less fortunate neighbours, a fact that he was not slow to appreciate.

The repercussions of the depression were not confined to farmers, for many landlords also suffered a loss of economic and social status. Many landowners derived their income almost solely from the letting of agricultural land, so remissions and, ultimately, reductions of rents in the 1880s and 1890s reduced their incomes considerably. As a result, some smaller landowners with no alternative income (i.e. from industrial or urban rents and investments) found their rent rolls so drastically reduced that the upkeep of their country houses and sporting estates became a luxury they could no longer afford. Consequently, these estates were either sold or let, and the close bonds of the landed interest, the traditional alliance of local landowning and farming families, was loosened as new owners or tenants moved in to the country houses and the old families moved out.

So the mechanisms that had controlled rural society for centuries were subtly altered and realigned while Calvertt was writing his diary. He saw

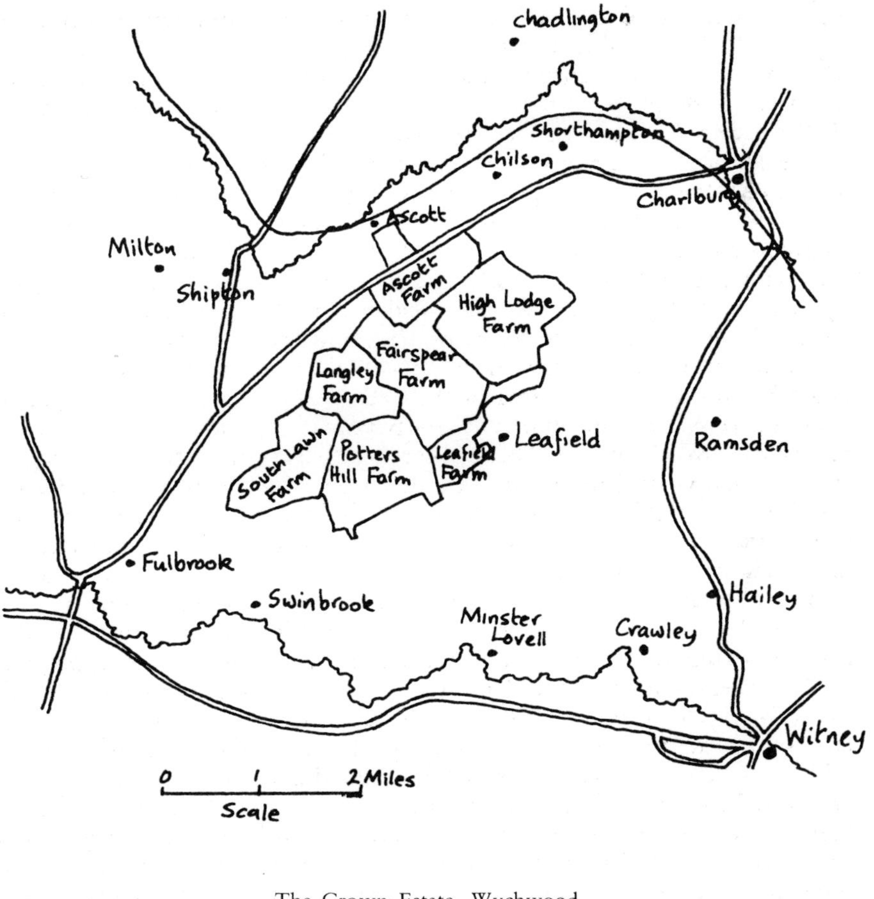

The Crown Estate, Wychwood

and recorded the changes and, whilst not always appreciating their full implications, he was only too well aware that the environment in which he was growing old was a very different one from that in which he had grown to maturity. We can see all this and more through his eyes. Calvertt was very interested in the politics of the day and confided his view of important events and his opinions of the personalities involved to the pages of his diary: a total veneration of Disraeli,and a grudging admiration for Gladstone, tinged with regret for his inevitable shipwreck on the rocks of Irish Home Rule. The growth of British imperialism is also reflected in the incidence of conflicts recounted: the Afghan and Zulu wars of 1879, wars in South Africa, Egypt and the Sudan in the 1880s, and the outbreak of the second Boer War in 1899.

This is the larger picture seen in the pages of the diary, but there is a smaller picture – denser in texture and much more detailed. It is a picture of the daily life of a well-to-do farmer of the period and his family: the sort of life they led, their place on the social ladder, their hopes and fears, expectations and aspirations, adventures and misadventures. In all, there were seven Calvertt children: Alice Louisa Jane (1855), Elizabeth Anne (1857),. John Charles (1860), Robert William (1863), Grace Catherine (1868), Frederick George (1873), and Isabel (1876), the only child born at Fairspear House. The two eldest girls married and followed their husbands to the far corners of the Empire: India and Belize. The two eldest boys followed their father into farming, while Frederick George moved to Coventry to work for Rudge and Co., the cycle manufacturers. Grace married into a local family of farmers and millers, the Matthews of Fifield and Shipton, while Isabel remained single. Calvertt himself was a sporting farmer, an expert and enthusiastic horseman and a keen hunt member and supporter, as well as being a noted game shot – all these activities were zealously recorded in some detail. But Calvertt was also a practical farmer and, although he employed a working bailiff at Fairspear, he involved himself in the life and work of his farm at all levels. His management of the farming business was acute and he managed to remain financially sound even when others around him were in difficulties. He also took a great interest in the labour he employed and the diary is a valuable record of agricultural wages and conditions of employment in the Wychwood area.

There is room for a only a brief and, I hope, representative selection of this voluminous record of the life and times of John Simpson Calvertt. For this is his story and he emerges from the pages of his journal – an individual, 'warts and all'. He is not always a sympathetic character to twentieth century eyes, but therein lies his virtue. He was a Victorian without apologies and without artifice. His innermost thoughts and emotions were rarely confided in the pages of the diary, but even so, the man stands before us. A confident, forthright John Bull, the backbone of Victoria's Empire.

Rain and Ruin
Extracts from the Diary of John Simpson Calvertt 1875–1900

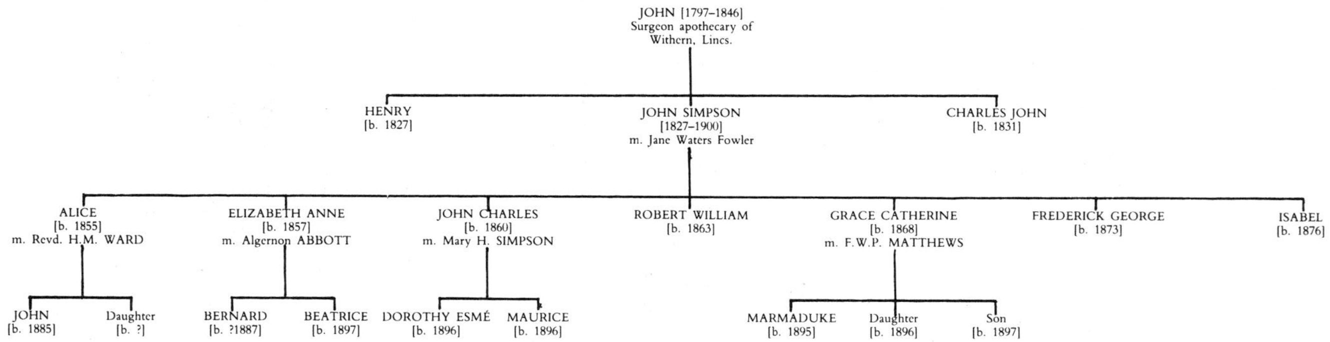

JOHN SIMPSON CALVERTT'S FAMILY

1875

Sept 23. Received a letter from F.E. Higdon[1] to say 'Fairspear', Mr. Heanley's Farm in Oxfordshire, is at liberty – I have asked for the offer. . . .

Sept 28. Walked to 9 Whitehall Place[2] and saw Mr. Clutton jnr. who instructed me about seeing Fairspear Farm – left at 2.45 the Paddington Stan. and ran to Reading & Oxford, changed and on the West Midland to Ascott, reached Mr. Heanley's Fairspear about 5.30 p.m.

Sept 29. Looked round the House, Buildings, & Land, & stayed the night.

Sept 30. Left Charlbury Stan. at 11.28 for Paddington and Whitehall Place – Saw Messrs. Southwell and Clutton and telegraphed for Mr. Heanley to-morrow.

Oct 1. Up at 6, walked to Farringdon St. Stan. – and on to King's Cross – hence to Finchley – saw John Chas. and Robt.Wm[3] who both acknowledged they would like to pursue farming thro' life – left and at 9 breakfasted at the 'Salisbury'.

Mr. John Heanley joined me at 10.45 and we left and reached Whitehall Place at 12: where I agreed to take Mr. Heanley's Farm – signed an agreement to that effect – we returned to [the] 'Salisbury' and Dined – Mr. Heanley left for Home at 6 p.m.

Saw Mr. Irving and Miss Bateman (Mrs. Crow) perform 'Macbeth' at the Lyceum.

Spent ½ hour each at 'Oxford' and 'Evans', was disappointed with both places.

Oct 2. Left King's Cross at 9 – stayed Peterboro' Fair and arranged my Land Sales &c.

Reached Claythorpe at 6.30. Tea'd at Mrs. Robson's and Home at 7.40 – after one of the most various week's employment about Business – I have ever passed or ever shall witness in the Future of my Life.

Oct 7. . . . A Telegram reached Tothill, to say Mr. John Heanley's Fairspear Lease was assigned by him to me, and accepted by the Commissioners of Woods & Forests.

So Fairspear will be our Future Home.[4]

Oct 11. By 6.57 train direct to King's Cross at 12 – by the underground rail to Paddington, left at 2.15 ran to Reading, Oxford, and changed to West Midland, and reached Ascott Station at 5 p.m. for Fairspear. Drove round with Mr. J. Heanley – and commenced my tenancy with Her Majesty.

Oct 12. Farmed round Langley after Breakfast.

Mr. Heanley's Stock Sale commenced at 1 p.m. Bought 16 of the best calves – was introduced to Mr. Savidge, Price, and Sergeant – Mr. Savidge was formerly from Lincoln Heath.

Very pleased with Mr. Jonas Paxton as auctioneer, the Sale was a nice introduction for me to all the Farmers, and others, of the Locality.

[Calvertt returned to Tothill on 13 October to supervise the sale of his stock and the parcels of land he owned, and for the reassignment of his Tothill and Grimoldby tenancies. On 12 November Calvertt travelled to Fairspear with his wife, Jane.]

Nov 13. My 46th. birthday.

Mr. J. Heanley and I met Mr. Upton, from 8 Whitehall Place at the White Hart Charlbury: where we signed the original Lease and finally completed the assignment of the Lease, for the next 17 years: We drove by Witney back, and called and saw Mr. Bradstreet at the London and County Bank, and brought away a blank cheque Book.

Very rainy afternoon and night.

Nov 14. Floods out all along the Valley of the Thames &c. from London, to Warwickshire.

The water in the Thames, London, rose higher than *ever* known.

Nov 15. Drove round with Mr. Heanley and met Mr. Moody at Langley. Met Mr. Jonas Paxton at the 'Churchill Arms', Ascott, at 2 p.m. and sold the Turnips on Fairspear, fetched £134 with 70 tons Clover, to consume with them on the land, and eat off, by 14th. Feby.[5]

Nov 21. Last Sunday attended Leafield Church adjoining Fairspear Farm – a newly built substantial church, with a spire and beautiful chancel, and good Organ, the Crown subscribed to its building in order that some of their Whichwood Tenants might make it their parish church – I like it and its situation very much – the Spire is a most auspicious object, in our Fairspear landscape.

[The slow process of moving from one farm to the other lasted until early April the following year, when the entire Calvertt family finally moved in to Fairspear House.]

1876

April 4. Rode to Station (Ascott) before Breakfast and Telegraphed Thomas Smith Esq., Louth.

Rode hence to Witney, Standlake, Kingston &c. – 18 miles to Abingdon – stayed night at Queen's Hotel (Pilgrim's) and Bought a new Hat, and sold my old one!!!

April 5. Left Abingdon at 5¼ a.m. reached Ilsley (11 miles off) at 6½ o'clock – Breakfasted at the 'Crown and Horns'.

Bought 60 Tegs at 49s. 80 at 50s. £2 luck, 200 at 57s. and 230 at 55s. – expended £1546.10.0.[6]

Mr. John Widdows, Standlake, was the *only man* I knew till I reached Fairspear: Left Ilsley at 11½ for Abingdon where I had Lunch and rode back to Fairspear at 5 p.m., very fortunate in having fine *fair* weather.

Good show, of Hampshire Downs and Half-breds.

April 6. Settled all up with Mr. Heanley and he drove away at 2 p.m. for Alton's, Rusper, Sussex.

My Wife, Alice, Elizabeth, Grace and Fred. reached Ascott, at 5 p.m. none the worse for their Journey.[7]

April 8. 4 tons 16 cwt 2 qrs weight of Furniture came *162 miles* by rail from Claythorpe via Sandy and Oxford, here: we carried it up, out of Box Trucks, and found it only slightly damaged.

Fine, dry weather – Drilling Spring corn and small seeds. I like the Labourers and understand their talk, much better than I expected.

April 9. . . . We all spent our first Sunday quietly at Home, and a great relief I found it – after one of the *very busiest weeks*, of my Life.

April 13. 4 inches of snow, on the level, at 6 a.m.: with high wind, and continued snow squalls during the day.

Met the boys at Oxford, on their way Home for Easter.

April 14. Terrific wind, & heavy fall of snow – Good Friday – *rough* Friday, & no mistake.

April 15. Snow melting rapidly. . . . Wonderful change in the weather, fine April day.

April 21. Mr. Mrs. Savidge & Mr. Abraham tea'd, and spent the evening. Mr. A. is an acquaintance, of the famous preacher, Spurgeon.[8]

Three horsemen left, because I would not give them 15s. instead of 14s. per week.

April 25. Carried the hay from Langley to Home yard. Drove to Mr. Savidge's & accd. him to Mr. Geo. Garne's, Churchill Heath, & witnessed a Short-horned Sale of 47 head – averaged £44.12.0.

Earl Fitzhardinge was Chairman, supported by Lord Moreton (Earl Ducie's eldest son) & a large company.

Tea'd at Mr. Savidge's – John Chas. accompanied me.

April 28. Marking 100 Best Ewe Tegs – placed 300 Tegs in Shakernook, for feeding ½ lb. Cotton Cake & Maize – 269 Tegs in Cockshott. . . .

April 29. Rode to Burford Fair – bought 5 steers £17 each and saw Mr. Hemming the Glazier &c.

Very showery – finished spring corn planting.

The weather remains most adverse for farming: and Lincolnshire is much worse than here.

May 2. Rode over to Chipping Norton Fair – Mr. Hitchman, had his annual Stock Sale, in his Brewery Yard.

Mr. Savidge introduced me to Messrs. Keck & Bickerstaff.

May 5. Dr. and Miss Batt, Witney: called.

Carrying off the stones, from Hall's Hill Red Clover, and repairing the low road to Ash Bottom, with them.

May 8. Turned 40 to 50 of the Beast out to grass in *daytime*: it is better they lay up at night, the first week, or fortnight.

May 11. The Prince of Wales safely reached Marlborough House, after a *magnificent* Tour of *Visits* in India, Egypt, Malta, Gibraltar, Spain, and Portugal, extending over *seven* months.

I walked out after Tea, for the *First time* with my Gun, since I resided here, to fire a Salute in Honour of my future Landlord's safe return – shot 5 times, and killed [a] Leveret, and 4 Rabbits.

May 28. Miss M. Fowler, & Elizabeth Anne, accd. me to Ascott Church, the new Minister, Revd. York preached, it being only his second Sunday's Duty.

Jane, Alice and Grace, accd. me to Leafield Church in the afternoon to hear *our* Clergyman, the Revd. Lee.

Miss M. Fowler, & Elizh. Anne, accompanied me to Witney Church in the Evening – Music, Singing, & Intoning, nicely performed, the Rector is a fine Apostolic Old Gentleman, called [left blank]: we returned back by Crawley, & saw the Blanket Manufactories on the Windrush.

May 29. Steam cultivation com[menced] in Shipton Field – did it twice over, in beautiful dry weather – Rolls, and Wholey's Harrows combed it over beautifully – Steamer came away on Thursday (9 a.m.) to Kingswood Walk, 110 acres.

June 1. Drove Alice & Grace to Witney: saw Mr. John Blake.
Returned Home by 5 Ash Bottom, and let Labourers, 120 acres Clover cutting – 31 Sainfoin & 9 Meadow; 3s.6d. all round, no Ale.[9]

June 3. Finished clipping 567 Tegs – & mowing 31 acres, Sainfoin, and thatched Tank House.
Light showers, & very growing.

June 4. Attended Leafield Church, in a.m., and stayed Sacrament.
Drove Alice & Elizh. to Shipton in Evening, & heard Revd. Baxter – service performed as Stylish as at St. Paul's.

June 6. Drove Alice & Elizh. to Mr. Savidge's, Sarsden Lodge – all attended concert at Churchill – except Mr. S. & myself, we talked & farmed round Home in the Evening – Supped, & into Bed at Fairspear, 1.25 a.m.

June 7. Drove Alice & Elizh. to Chipping Norton Fair – attended Board Guardians – Dined White Hart, & Home at 3.45.[10]
Shot 3 rabbits in evening.
20 acres Turnips in, in Shipton Piece.

June 9. Leafield Holiday to-day & to-morrow: scarcely any work done till next Monday – quite the Custom.[11]
Cultivating 'Kingswood', 2nd time over.
Manured Boynal Gardens 9 acres, & pt. of Leafield Close.

June 10. Mr. Beecham, from the Crown Office, inspected Kingstanding Spring, Spouts and Tank – for laying water on to the whole of the Crews etc.
Leafield 5 Years Club, instituted in 1784, called and played a selection of music – gave them a Bottle of Sherry and 5s.[12]

June 14. Mr. Robert Hambidge, rode all round, in afternoon, & stayed Tea – told me how he had been persecuted by the Josh. Arch-ites, 2 or 3 years ago.[13]

June 17. Commenced digging out, a substantial 2 inch diameter pipe, in Far Lawn, about 100 yards long – it will be very useful.

The Steam Cultivator finished *twice* over Shipton Piece 47 acres, Kingswood 110 acres, & 26 acres of the Farthings, and very good work it has made – consumed 34 Tons Coal, Water leading, & paid the Oxford Co. £140. . . .[14]

June 19. 'Apology' won the Gold Cup at Ascot last Wednesday – 'Kisber' has won the Derby & Grand Prix at Paris – both were descended from the Revd. J.W. King, Ashby de la Launde, Lincolnshire – and Sir John Dugdale Astley's 'Hopbloom' won the Royal Hunt Cup at Ascot – so well done my dear Old Lincolnshire, Racing men, & Horses.

Mr. John, and Miss Savidge, drove over here, and spent the evening with us.

June 26. 12 scythes cutting Clover – Hall's Hill 95 acres – and Langley Meadow: Turnips growing slowly: owing to the very bleaching weather: which exactly suits the Fallows.

June 27. Finished cutting, & clearing, all the Thorns, Briars, Clematis, etc., out of North end of Priest Grove, intend to add it to the 7 acres arable, & graze it with Horses next Summer – altogether it is a great improvement, and will [be] most useful, instead of useless. . . .

June 28. Met Annie W. Calvertt, at Ascott 5 p.m.[15] – came here, after visiting in Surrey, to spend 2 or 3 days, before she returned into Lincolnshire.

June 30. The 12 mowers finished all up – and have cut about 180 acres of Sainfoin, White & Red Clovers, & 8 acres Langley Meadow – 3s.6d. per acre all round.

Had a deal of *fine, dry,* weather with only *slight rains* during this month – but it is surprising the work we have done *on* the *Farm* with *30 Horses,* and upwards of *50 hands* – of all sizes: this will be, by far, my busiest spring at Fairspear.

July 2. Annie W. Calvertt, Ashby: taken ill, fainted at 3 p.m. I rode over and tea'd with Mr. Savidge, Mr. Sargent joined us, had a very pleasant evening – but on returning Home at 11 p.m. the groom (Sifford) had gone for Dr. Batt – Annie W. Calvertt having vomited lots [of] blood, and appeared almost in a dying state – the Dr. thought a blood vessel had broken in the stomach, thro' congestion.

July 3. Dr. Batt here at 11 a.m. and found Annie worse – and after he left she vomited dark blood and looked like dying.

Messrs. Clutton sent me rent audit form.

Carrying Clover out of Hall's Hill – 6 waggons.

Telegraphed for Mr. Robt. Calvertt, Ashby, nr. Spilsby.

July 4. At 11 a.m. Annie Calvertt had another attack of Vomiting Blood off the Stomach, only 5 minutes before Dr. Batt called – to our great relief!!!

Met 5 p.m. train but no Mr. Robt. Calvertt to my great regret – but he reached here, via Witney, 7½ p.m.

July 10. Fine, *sunny*, clear morning – comd. 1½ days Clover carrying.

Annie W. Calvertt: had no return of bleeding from the Stomach, and apparently going on most favourably.

July 11. Shot a Magpie and 3 Rabbits, in Priest Grove, 5½ a.m.

Finished Clover carrying, and very relieved, and pleased, I feel, having secured about 180 acres of Clovers, Sainfoin, and Meadow, in *Good Condition*: and probably with less expence, than I ever may expect again for years.

Mr. Shillingford, Witney, bought wool, yesterday, 33s. per tod.

July 13. Drove to Sarsden Lodge, & Mr. M. Savidge joined me for Hound Puppy judging & Luncheon at Chipping Norton Kennels – Mr. A. Brassey, A. Rushout, Sir M. Crofton, Messrs. Matthews, Mathers, Price, Hayward, Scovell & 80 others attended – very jovial party – left 7.30 and called for an hour, at Mr. Sargent's, Churchill – reached Home about 10 p.m.

July 14. Robert & Annie Calvertt left for London, by the 1.24 train where they would stay till noon to-morrow, and journey to Ashby, in the afternoon.

July 17. Wm. Sampson already thatched 5 Clovers, and 2 Hay and Sainfoin, Ricks 9d. square [foot] –

Had 4 or 5 very *sunny, Hot*, days.

Yesterday, drove over & Tea'd with Mr. Widdows, who is laid up, of an abscess, under his Knee.

Let 17 men, Shipton Piece, and Boynal Gardens, *Swedes*, at 10s. per acre – very weedy and difficult doing – *single out*, and then *Flat* Hoe *all* over –

2 men, commenced same in Leafield Piece, at 8s.6d. per acre – but they are better work – altho' very weedy. . . .[16]

July 18. Messrs. Shillingford came and took up Wool – 567 Teg fleeces weighed 137½ Tods, 33s. – Cheque £226.17.0.

Alice by 1.24 train to Huntingdon.

Jane, Eliz., and I, tea'd with Mr. Mrs. Abraham, Ringwood Farm, nr. Minster Lovell, and met the famous Mr. Spurgeon & a large concourse of Neighbours, & Friends, who listened very attentively to two addresses given by Mr. Spurgeon at 2.30 & 6.30 p.m.

Made the acquaintance of Messrs. Bliss, Gering, Spurgeon, and Blake, &c – Home at 10 p.m.[17]

July 19. By 8.55 train via Evesham to Birmingham – reached Royal Show at 1 p.m. – met with Mr. Blake, Savidge, and Wholey: bot. 2 skerries, Mangle, & ordered new Corn Drill for 1st Septr., of Messrs. Hornsby – stayed night at 'Stork Hotel' – Very successful show.

July 20. In show yard 9 a.m. had a long day and saw a great deal all over the yard – attended the Council meeting at 3½ p.m. Mr. Stillgo, obtained the £100 Prize, for the Best Cultivated Farm – to the great delight of Mr. Savidge, & his Friends.

Lord Skelmersdale, Evesham, Leigh, Moreton, Sir Watkin Wms. Wynne, &c: attended the Council meeting.[18]

Left Snow Hill Stan. at 5.12 and returned by Stratford and Honeybourne to Shipton Stan. at 8.12 p.m. very pleased with my visit, and all I saw.

July 21. Drove, stripped in shirt sleeves, & umbrella up, to Witney – making 10 sunny, *hot*, consecutive days.

Mouth and Tongue, sore.

July 25. Drove Jane, & Grace, via Leafield, Finstock, Fawler, Stonesfield, to Bear Inn, Woodstock; walked over the Park and Gardens and inspected Paintings &c – in Blenheim Palace – returned by the Palace, over the Bridge & across the Park to Mr. Gering's, saw Mrs. G. & came away thro' Black's Lodge, Home at 5.45 p.m.

Mrs. Browning, Milton, and Oxford, called.

Very sunny, and hot.

Saw Kangaroos, & Emus, in same *Paddock*, in Blenheim Palace, strange to say, they *eat Indian Corn & same food together.*

Last evening Mr. Savidge, & Sargent, tea'd: & farmed me round, & bot. a Reaper each of me – also recommended a Windsor customer, for 6 of my Bullocks, & 100 wethers.

July 29. Started W. Smith's, Kettering, 2 new Turnip Hoes, bot. of him at Birmingham – they work admirably.

Rode over to Burford, in evening, & saw Mr. East. . . .

All ploughed (except 4 acres): Swedes singled, in Shipton Field, & Boynal Gardens; *all* Manure lead out, at Home Yard, 30 acres Leafield Turnips 'singled' and at last have got fairly forward with *all* the work and shall be nicely ready for Harvest.

July 31. Had a grand burn, & clean up, at Kingstanding Yard, of all the refuse chaff &c.

Townsend came from Curbridge, & feasted the Rats for poisoning on Wednesday.

This has been, a very dry month, throughout.

Aug 2. Drove John Chas. & Eliz. Anne to Chipping Norton Fair, they looked all round the place, and joined me at Dessert, after I had dined with Mr. Bliss, the great Tweed, & Shawl, Manufacturer. Mr. Mrs. Garne & Mrs. Garne, (Charlbury) were also there, we all inspected the Manufactory, Dairy Farm, & 9 Shorthorns – walked into the Wilderness, & up the Lovers' Walk, &c – when I left, a Shawl was presented to me by Mr. Bliss for Jane – Home at 9½ p.m. after spending a most pleasant day.[19]

Aug 9. By 8.22 train for Charlbury, & attended the Great Ram Fair at Oxford – Mr. John Savidge accd. me an hour round the City, admiring the Colleges.

Dined at the Cross, with Messrs. Savidge, Stillgoe, Garne, &c.

Had wine with Mr. Blake – left 4.20 – very *hot.*

Aug 10. Began cutting Oats in 'Calvertt's Corner' with 2 of Messrs. Hornsby's 6 arm Reapers – they do the work admirably well – and are very light in 'Draft'.

One man began 'Fagging' in Priest Grove.

Aug 12. Nearly all hands Harvesting – weather *hot*, and calm, 2 Reapers & 7 men Fagging at 10s. to 8s.6d. per acre.

Grace Catherine's 8th. anniversary, gave her, Her Majesty's likeness in Silver, 1s.

Aug 15. Rode, and called at Cornbury Park House – but found Lady Churchill was at 5 Bury St., St. James, London S.W. so came away, and took the Old Butler's advice, and wrote to her Ladyship.[20]

Going on very satisfactory with Harvest operations – 1 Mower, 12 Faggers, and 2 Reapers going, upwards of 200 acres cut.

Aug 17. Drove John Chas. & Robt. Wm., & Alice, to Witney: very dusted, & nearly roasted – weather *is*, & has *been, for weeks*, very sunny, & hot.

Carried Langley 12 acres Barley, Priest Grove and Woefield Green Wheat, the shocks under 10 Oaks in 'Calvertt's Corner', and part of the Farthings Wheat – a good day's carrying for my first in Oxfordshire.

The 2 six-rakes Reapers, go well, in Wheats, Barley, & Oats: and have quite answered my expectations in every way, & *a great help*, and *relief* I find them, at this season.

Aug 18. A shower at 4½ a.m. and gentle rain from 10 to 12.

Lady Churchill sent a letter, by her Keeper, Shayler, regretting injury I had sustained by her rabbits, and would take care they are not injurious to my crops, in future: also offered to take Fairspear shooting!!??

Drove Alice & Elizh. to Mr. Abraham's, Ringwood House.

Aug 19. Awful Thunder & Lightning, set in, with *heavy* rain, at 3 p.m. – the finest rain, and *first* Thunder, I have ever experienced at Fairspear.

Aug 22. John Chas. & I, by 8.55 train to Stratford-on-Avon, there met by Mr. Wm. Hutton and driven by Sir C. Mordaunt's, & Lord Leigh's Parks, to Compton Verney, looked over the House, Gardens Stables, &c: and were joined at 6 p.m. by Mr. Edwin Tickler, at Dinner – spent a pleasant evening.

Aug 23. Up at 6 a.m. walked round the Home Farm, the Oat & Barley crops, were especially good, & heavy. After Breakfast Mr. Hutton drove us to Kineton, saw Mr. Malcolm's House &c – Lady Willoughby's Kennels – and the Battle Farm.

Lord Willoughby [de Broke] took Mastership of Hounds on 6th April – so follows in his Father's steps.

Luncheoned at Mr. Hutton's, and we took the train at Kineton for Stratford, Warwick and Leamington. Mr. Hutton left for London, & we retd. to Shipton at 8.12 p.m. after enjoying the visit very much.

Sept 2. Carried Harvest Home.

Gave 1s. per acre for 341a. 2r. 0p. Reapered, 8s.6d. to 10s. per acre for 91a. 3r. 0p. Fagged, and 3s. per acre for it all carrying and stacking.

John Chas., & Robt. Wm., walked out with me in the afternoon, and I shot brace Leverets and Partridge in Hall's Hill, a Partridge in Leafield Piece, a Rabbit in Langley Grove, and another Partridge in Farthings: very unwell of Diarrhœa, so came into House: saw a nice lot of Birds.

Settled *all* up with Harvest men: and Labourers: and feel very pleased to have secured Harvest with so little trouble & no extra expence whatever, thro' wet weather: all thatched at Home Yard.

Sept 3. Walked round with John Chas., & Robt. Wm., beautiful sultry day for Turnips, Clovers, &c: distant objects look near and the wildness of Partridges yesterday, betokens a change in the weather.

I rode over and Tea'd with Mr. Stillgoe, at Mr. Savidge's, Sarsden Lodge – who has Barley, Oats, & Beans, to carry yet: and the *worst* Harvest he ever grew – in quantity of straw and yield of corn.

Sept 4. Rain from 4 a.m. till 9.

Pastured 'Calvertt's Corner' with the 18 Langley Bullocks during the day, & back to Water, & their grass at night.[21]

Have named the 'Big Field', 'Calvertt's Corner' because it is not the biggest field – and my other name is more amusing.

Stood an hour under Kingswood Clump, and shot 14 Woodpigeons.

Heavy rain at night – just suits my young Clovers, Grass, & Turnips –!! a *great deal* of Corn out *20 miles around* here.

Sept 8. Drove John Chas. to Chipping Norton Agricultural Show – Dined at the White Hart – A. Brassey Esq. – in Chair – Earl Redesdale, Col. North, Mr. Bliss, Reed, &c, &c.

Sept 9. Rode to Burford – neither saw East, nor Jacobs, but caught Mr. Samprey at Home, at Fulbrook.

Settled all up for Thatching – Measuring – and entirely finished up a/c's & expences, of this, my *First Oxfordshire Harvest.*

Hired under Shepherd – Edmd. Hopkins: House & Garden, 120 Faggots, £3, & sack malt, & 15s. weekly.[22]

Sept 14. Mr. Samprey came and inspected the Steam Engine, and Thrashing part.

Drove John Chas. & Robt. Wm. to Witney: bought some Figs, & Prunes, to accompany Robt. Wm. to school on Monday next at Christ's College, Finchley.

Very showery in p.m.

Sept 15. Had my first day's shooting from Home in Oxon., at Mr. Abraham's, Minster Lovell: Mr. Akeman, Witney met me – we killed 11 Birds, & 6 Hares, the game *very* wild – Ringwood is a very nice Farm for shooting, but hungry Gravel, spoils its utilisation for cereals.

Sept 23. Showery morning.

By 1.24 train to Oxford – hired Josh. Shepherd as Carter – Took for Charlie, 36 Woodpigeons, to Mr. J. Turrill's.

Sept 25. John Chas. accompanied me to Burford Fair – hired John Carter for Shepherd, & 2 married, & 1 single, Carters.

Sept 27. Rode with Mr. Wm. Abraham, by Chilson, Chadlington, Dean, and forward to Heythrop – Lunched with Mr. Blake, and we walked across the fine old well-wooded Park, and by north end of Heythrop Hall, to Mr. Geo. Abraham's Little Tew, and attended his Stock, & Implement Sale: tea'd: and we retd. on foot over the Park to Mr. Blake's – reached Home at 10 p.m.

Saw some good, and some very bad, farming: especially in the grass valley, of Dean.

Sept 29. Sat in Ascott Common Clump from 9.30 a.m. till 1.30 p.m. and shot 43 Woodpigeons – made some of them at great distances: to see how far off the Gun, would kill them.

Levelled the old House foundations at Langley – Leading stones off Cockshott Hill into Forest Road, and thorns out of Ascott Road.

Very fine breezy day, after a deal of rain.

Sept 30. Very boisterous rainy day – all day long!!

Mr. Dring came to my Black Horse, dreadfully swollen about the Head and Neck, thro' screwing himself up, by entangling his hind foot, in Head-stall, while running loose during the night – if he dies, his end will be suffocation. . . .

Oct 3. My Black Horse died – from his accident in Headstall – very temperate nice quiet Horse in every respect, and perfectly sound – ride – drive – hunt – very bad luck!!!

Tea'd at Mr. Price's – & called at Mr. Josh. Reade's, Shipton.

Mild, fine, evening.

Oct 4. Drove Eliz. & John Chas. to Witney Fair & Statute, or *Mop*. . . .[23]

John Chas. and I, dined for *our first time*, at the Marlbro'.

Oct 9. Sold an old Horse, £6.15.6, & a giddy Lamb 27s.6d. at Shipton Auction.[24]

John Chas. & I, tea'd at Mr. Maddox.

Dores plashing hedge by Shipton Road, & leaving a few Hawthorn bushes, for ornament: 19 chains completed for £2 – which is cheap!!!

Oct 13. Drove to Mr. Josh. Reade's, Shipton, and he drove Mr. Geo. Garne, & I, to shoot over Swinbrook – very *showery* only bagged 6 Birds, 3 Hares, & 1 Rab:

Mr. Savidge joined us at a Venison Dinner, Dr. Hague joined us at 9 p.m.

Mr. Reade lent Charlie his only Son's Gun.

Oct 14. Dipped 317 lambs – by Martin.

Very rainy, showery week: Wheat sowing, sadly *slow*!!

2 Misses Hambidge, to Tea with Grace.

Oct 18. Off at 4 a.m. – to Standlake for Breakfast & accd. Mr. Widdows to Ilsley, bot. 98 Down Theaves 48s. – we returned by Wantage – there saw a Bear erected in front of Inn with a bunch of Grapes in his mouth – looked over the Old Church &c. Tea'd at Standlake and to Fairspear 8.30 p.m.

Mr. Aldworth, Frilford: was ploughing a Clover field, sown with rags!!![25]

Oct 19. Firm Corn trade at Witney market – begin to feel quite at Home, and know all the principal merchants &c.

This is the 4th fine day!!!!!

Oct 20. Went out, or rather up, to High Lodge, and cleared out the wretchedly conditioned Privy – to the great surprise of the 2 Ascott, Mosses: who required Brandy, Gin, Tobacco, to enable them to bear the aroma!!!

The 98 Down Theaves reached here in nice trim, after their walk from Ilsley.

Oct 21. Heythrop Hounds made their first call here on their way to Worsham Mill nr. Minster, for Cub-hunting.

John Chas. shot his second Hare, in same field as on 1st Sept. ult., Shakernook.

Dores *over-fed* the chaff-cutter, and the wheel broke.

G. Ide, walked at Lillie Bridge, last week, 50 miles, in 8 hours 19 minutes 55 seconds!!! being the fastest walk, over the distance, on *record.*[26]

Oct 23. Dores, commenced plashing, & trimming up, all around Highlodge.

Mr. Price tea'd with us, & spent the evening, & gave a dissipated account of Sir Morgan Crofton, who now resides at 'Barton Court', Shipton-under-Wychwood.

Oct 24. Rode thro' the Farm, & over Mr. Morris' to Chilson, across the fields & thro' Chadlington, up thro' Dean Valley and over the Hill to Lidstone, and Heythrop; met with Mr. Blake, and rode down the Park to Schools and new Lodge – saw 350 Oxford Down Ewes – met with Mr. Brassey shooting a few Ck. pheasants for next *Monday's Lunch* being the *first* public *meet* of the Season, at Heythrop Hall – a 1000 are expected – !!! – we rode on and inspected 50 Highland Scots [cattle] lately bot. in the North, 4 Shetland Ponies, & a Colley Dog – rode all around the House, & returned to Lunch at 2. Left at 3, & came same route back – went to look at a 4 years' old Black Mare, but she did not suit me: called at Kingstanding & Home at 5 – after enjoying the day very much!!!

Oct 25. By 8.15 train to Oxford – saw the magnificent painting by Long, called the 'Pool of Bethesda' sold for £3000 – looked over a Poultry Show and attended a sale of Horses in the Clarendon yard, and Dined with Mr. Blake, Somertown: reached Charlbury 8.40 p.m. after spending a long, *pleasant,* day in *Grand, Old, Oxford.*

Oct 30. Started 4 drags on the far-side-half of Calvertt's Corner Fallows, ploughed 4 to 6 weeks ago – makes a capital tile,[27] this dry weather!!!

Drove Alice and Eliz. to the grand annual 'Meet' at Heythrop Hall – we Champagned – saw over part of the House – the Stables – Gardens – &c. They killed a Fox near the Ovens Gorse, and came down the Park, and over the Iron Bridge & soon found again, & ran out for Tew, & Rollright.

We Luncheoned again at Mr. Blake's – with Mr. Mrs. Stephens, Irving, and Savidge: & looked the yards over &c – reached here 6 p.m. after a very pleasant day's 'outing'.

Supposed to be 1000 persons, witness the Opening, it has been an *annual custom* for *years* to give an entertainment to *all comers*, on opening the season

– Mr. Stephens said he had never missed attending during the past *31 years.* Saw 3, of the 4 young Brasseys, in the Park: their birth-days are all in *this* month.[28]

Nov 1. Attended Mr. Iorns Furniture Sale, & sent some of Mr. Heanley's Furniture, and brought away a Cart, Wheelbarrow, & Ladder £7.19.3d. bot. yesterday.

Fine dry cold morning with first *black* frost, ice over the Langley Ponds. Messrs. Paxton, Son, & Castle, *all* assisted, in selling at South Lawn. Thrashing, & 4 Drags crossing Fallows!!! daily.

Nov 3. Attended Leafield School meeting – Revd. Lee, Abraham, Bateman, Edelston, & myself – agreed to pay 3d. rate.[29]

Nov 4. By rail to Oxford market. Sold Mr. Birmingham 90 qr. Barley 36s. – and Mr. Butt & Skurry 23 at 30s. Dined at the Roebuck Hotel, Mr. Castle in the chair – also 55½ qrs. Wheat to Mr. Gomm, at 48s.

Alice, Eliz., & John were driven to meet Hounds at Blenheim, and were very pleased with all they witnessed – one of the most stylish 'meets' in England!!!!

Drag-harrowed twice, half Calvertt's Corner, & Boynall Cover piece, 95 acres – Fallows for Turnips ploughed early!!!

Nov 9. Alice accd. me to Witney in p.m. – Home at 5.30.

Witney market for Sale of Corn from 2 to 4, makes it most convenient for us, as there is time for a good day at Home, leave here at 1, and here again for Tea at 5 to 6 o'clock.

Nov 11. At Oxford – Mr. Gomm 71 qrs. Barley 36s. and 130 qrs. 32s. To-day makes 12 days, thrashing Wheat & Barley, without an hour's delay, by Wind, or Rain!!!

Hounds in Priest Grove – Reynard went to Ground, regretted I was away.

Mr. Edgington very ill – of Churchill: one of the old school of Farmers.

Nov 12. Cold wind and rain – after *5* weeks of comparatively very fine, *dry*, weather.

Capt. Nares, & Stephenson, returned, in the Alert & Discovery, after 18 months absence, attempting to reach the North Pole, they advanced to within 400 miles – but ice, and severe cold, stopped all further advance: and all has been done, that human courage, & energy, can accomplish.

Nov 13. The terrific wind of Saturday night, has very much broken the branches on east side of the fine Oak, near the room window.

Mr. D. Young, Witney: came and inspected the Elevator and Thrashing Machine.

Misty morning, with rain all the afternoon, and evening.

Nov 14. Contracted my men, and work, so that I am now able to walk round my Farm.

Dined at Mr. Geo. Garne's, Churchill Heath, and met Mr. Blake, McIver, Mrs. Cheetle, &c. Home at *4 a.m.*.

Called at Mr. Savidge's and they reported Mr. Edgington dangerously ill.

Nov 16. Henry Chaplin Esq., M.P. Blankney Lincolnshire: yesterday married the Duke of Sutherland's daughter, the Lady Florence Leveson-Gower.[30]

At Witney, saw a display of roots grown by J.B. Lawes' manure.[31]

D. Young repairing my Thrashing Machine and Elevator.

Nov 20. Killed 3 Berkshire Pigs: Burnt, instead of scalded them, the opening, and hang-up style, is very different to Lincs.[32]

Nov 21. A 4th daughter born – Isabel.

Nov 25. By 11.28 to Oxford – Recd. £662.11.0.

Luncheoned with Mr. Oliver at Queen's College and looked round the Chapel, & Library.

Dec 2. Mr. Price called and I rode with him and met Hounds at Cornbury Park – unfortunately a favourite Hound died at the 'Meet' apparently poisoned!!! Mr. Brassey said he dare not draw the covers, & we went away, & found at Wilcote – lost at Stonesfield, found at Singe Wood, killed. Rain: I reached Home at 2.15 – Hounds went on to Stockley: the Hound's death, quite marred the pleasure of the day.

Dec 5. Alice & I, by 8.14 train to Paddington & the Salisbury Hotel and entered the Ag[ricultural] Hall, Islington at 12.45. Bot. 2 Cutters £19.14.6 and £4.17.6, settled with Mr. Smith, Wholley, and Hornsby.

Very good show, of very good animals.

Saw several of my Lincolnshire Friends, viz Cartwrights 3, Antons 2, Heanley, Martin, Davy, Hill Johnson, &c –

Met Charles at Hotel.[33]

Witnessed performance of Richard III by Barry Sullivan, at Drury Lane – acted magnificently!!!

Dec 6. Walked to Ridgway's before Breakfast and bot. 7 lbs. Tea, & 6 lbs. Coffee – £1.10.6 – 3s.6d. & 1s.8d. per lb.

Alice & I, walked to Blk. Friars Bridge, Mansion House, Royal Exchange, Bank, and by 'Bus, to Regent St., bot. a Mustard Pot at No. 166: 26s. – walked thro' Bazaar, Oxford St.

Christy's Minstrels at 3 p.m.

Witnessed Sanger's magnificent performance of 'Waterloo' and the riding of Miss Sanger, and 3 children: &c.[34]

Dec 7. Rainy morning.

Drove along Thames Embankment to new Ho. of Parl., Westminster Abbey, & Aquarium, looked all well over and to Maskelyne & Cook's Seance at 3 p.m.[35]

Witnessed 'Jane Shore' at Princesses.

Dec 8. Spent 2 hours at Madme. Tussaud's, at 2.15 left Paddn. Stan. & reached Fairspear at 5½ p.m. – after a very agreeable visit in Town.

Dec 9. Drove Alice, Eliz. and John Chas. to Dytchley Park & saw the Hall, Deer, & Meet of Mr. Brassey, and his Hounds, drove about after them for 1½ hour, to Glympton Covers, & back to Ditchley shrubberies – Home at 2 p.m.

Dec 18. Lord Redesdale, Batsford Park, Gazetted Earl Redesdale, County of Northumberland: for his past services in the House of Lords, he is Chairman, and upwards of 70 years old.

Dec 19. Mr. Price called, & I accd. him to meet Hounds at Barton Lodge – we called upon Mr. Rowland, Wootton.

Saw Mr. Barnett's Park & House, at Glympton, & Mr. Gaskell's Home at Kiddington.

Dec 23. Frost with slight snow – manured the Farthings.

Shot Hare & couple Rabs, in Boynal Cover – saw 2 Foxes on the Farm yesterday, & one to-day.

Our baby girl, born on 21st. November, named Isabel – on Monday last: at Leafield.

Dec 25. Spent to-day as quietly here as at Tothill last year, and what changing scenes I have passed thro' the last 12 months – probably never to be equalled again in my Life.

At Church in a.m.

Dec 26. Frost – Carrying out Manure – to keep Horses, and men, employed.

Very *cold, windling, sleety, snowy*, evening: which turned to drizzly rain at Midnight.

Shot 7 rabbits, and a Hare.

Dec 31. At Leafield Church in a.m.

Elizh. & Grace, attended Shipton Church, in evening, while I talked to Mr. Price.

Adieu 1876, a year of great Change to me!!!

1877

Jan 1. Rode thro' the floods at Ascott and called upon Mr. M. Savidge: who accd. me, & showed me Sarsden House, the future seat of Lord Moreton, Earl Ducie's eldest son: we joined the Hounds in Sarsgrove, blank: Churchill Heath, blank: over the Evenlode & floods, into Bruern Covers – found & ran to ground in 10 min. – all the rest blank – I came away Home – Hounds went for Tangley.

Alice and Elizh. accd. me to Dine at Mr. Reade's, Shipton; met Mr. Tucker, who sang the National Anthems of England, Germany, France, and Spain –

Mr. Read's only Son and Daughter, at Home, Holidaying.

Jan 3. Rain – Rain – the Valley of Ascott, Shipton, &c, looks like an inland Lake – Floods all over England & Scotland.

Jan 4. Drove by Crawley, thro' the water to Witney – where it was running thro' the Houses.

Accompd. Alice, Eliz., & Miss Read, to Witney, & attended Choral Society's Concert – in Corn Exchange – Mr. Habgood, sang best.

Jan 6. Rain,rain, & wind: *most wretched weather.*

Mrs. Thomas Brassey died ae. 66 – Mother of Thos., Henry and Albert Brassey – Hounds did not meet at Shipton Court, because it was her funeral day.

Jan 7. Frequent rain squalls – England scarcely ever was visited, with the weight of water which has flooded all the adjacent lands, & towns, thro' which the rivers &c pass.

The Queen spent Xmas at Windsor, and Prince Wales – at Sandringham.

Jan 8. Sold 20 Shear-hogs at Shipton Auction, they averaged 72s. each: 9d. lb.

Alice, Eliz., & Grace, with me in Waggonette to Witney Childrens' Ball – had Champagne with Dr. Batt, and Supper with Mr. Shillingford – became acquainted with Mr. Worley, Brize Norton, & Mr. Robt. Matthews, Church Handborough.

Jan 10. Met the Hounds at Bradwell Grove, ran a ring-ing Fox and lost him – found 2nd in Jolly's Gorse, in which the Hounds ran $1\frac{1}{4}$ hours and killed – found a 3rd & ran to Sherborne Cow Pasture – misty day – Mr. Price piloted me to Shipton.[1]

Jan 13. Sent 7 Bullocks, & 60 Shear-hogs from Charlbury for London on Monday – sold by Mr. Denchfield. They realised £400.7.6d. net: Beast £27.15.0d., Sheep 72s.

Shot 7 Woodpigeons.

Mild as April – birds whistling.

Jan 16. Beautifully still, mild morning – Violets, Primroses in bloom, & Thrushes & Blackbirds whistle like April.

Alice, accd. me to Witney Athaneum Soiree, we reached Home 6.45 a.m. after spending a very pleasant night.

Jan 28. Rainy morn – Alice, & Eliz., to Witney church.

Drove John Chas. and Robt. Wm. to Mr. Savidge, Sarsden, Tea'd & talked till 9 p.m.

Vivid Thunder, & Lightning, with Snow at *10 p.m.* – never saw the like in my Life – frost afterwards.

Jan 31. By 8.17 train to Oxford, attended Fat Stock Market & a Sale of 45 Horses at the Clarendon, sold by Mr. Paxton – Mr. Price bot. a grey 44 Guineas.

Mr. Turrill allowed John Chas. 1s.3d. each for Golden Plover, 1s. each for Grey Plover, 1s. each for Snipe, & six pence each for small Snipe.[2]

Feb 1. Lady Churchill sent me a brace of very fine-fed Ck. Pheasants.

Shot 20 Woodpigeons last Tuesday sat by the wall in Boynal Cover.

Alice, Eliz., & John Chas. accd. me to Witney.

Mr. Hambidge, spent the evening here.

Feb 2. Joined a party of about 40 persons, at Shipton Conservative Society – Mr. Morris in the Chair, Mr. Lambert in the vice-Chair – the Revd. Yorke made a very good speech – I responded for the visitors: enjoyed a very pleasant, instructive, evening – had cigar &c – with Mr. Reade afterwards, reached Home at 1 p.m.

Feb 3. 90 Tegs, 70 at 48s. & 20 at 35s. sent by Mr. Eagle – 30 Shearhogs off to London, 200 sold since 8th Jany.

Mr. Savidge rode over & shepherded his 250 ewes but did not stay.

Dores cutting and tieing up the underwood in Chalk Pit Spinney.

Weather very wet, & mild.

Feb 6. Misty morning – did not go to Taynton, to witness coursing at Mr. Matthews: they killed 14 Hares.

Alice and Miss Clara Oliver, sang at Leafield Concert – Mr. Sargent tea'd, & accompd. us.

Feb 11. John Chas. walked round with me, by High Lodge.
Alice, Eliz., & I, were driven to Churchill Church, the Revd. Carter, preached – a plain substantial edifice built by Sq. Langston – Churchill is a very pretty picturesque village.
We tea'd at Mr. Sargent's – The Mount.

Feb 12. Frequent showers, during the day –
Cutting out all the underwood &c in Chalk Pit Spinney – give 1s. score for its being cut and tied up into Faggots, which sell at 2s.6d. score, to Mr. Webb, the Baker, of Leafield.
Hurdles repaired for Lambing.
A nice show of Partridges on the Farm for breeding, only killed 30 brace, altogether, this season.

Feb 17. Beautiful cool breezy drying day – Met Hounds at Shipton Barrow – blank – but found in Widley, ran thro' the Covers & away by Mrs. Abraham's, Minster, to ground at Leafield, nr. the Lime Kiln – found again at Stockley and ran to ground in Swinbrook – tried the Beeches, Falsgrove, Furze grove, Langley Covers, Priest Grove, & found in my Boynal Cover, and ran across Ascott Common into the Forest – when I left.
Mr. Sartoris presented me with a Bouquet and Sir Morgan Crofton, & Mr. Price, called on returning Home.
Saw Mr. J. Reade, & altogether had a most agreeable day.

Feb 18. At Leafield Church in a.m.
John Chas walked with me to Mr. Price's by 3 p.m. – when we all walked to Shipton Court: Sir Morgan Crofton showed us his Horses: all over the House – saw a sword given by Lord Nelson, to one of his Ancestors, after the battle of Trafalgar – his Father's Medals, &c: of the Crimean War, where he was twice wounded – also some magnificent racing cups, won by himself, and a many articles of Vertu: &c: spent 3½ hours in the *old romantic House*, which was built in 1620 – and belongs to my friend J. Reade Esqre.
Tea'd at Mr. Robt. Price's.

Feb 21. Attended a meeting, and met Mr. Habgood & the Revd. Lee, at Leafield School-house, respecting a heating apparatus for the Church.
Mr. Mrs. John Iorns tea'd with us.

Feb 22. Commd. Oat threshing.
John Chas. accd. me to Witney. Ordered 8 Tons of Nitrate Soda, 8 Tons Salt, & 4 Tons Pure Linseed Cakes – 18s. – £12.15.0d. & 15s.10d., and £11.7.6d. per Ton.

Feb 25. Very high West wind. At Church in p.m.
Mr. Maddox & Abrahams, having great loss amongst their lambing ewes: &c.

Fredc. Geo walked with us to Leafield Church, for the first time – he said he liked the Piano (Organ).

Feb 27. Rode to Taynton and called on Mr. Matthews who showed me the road to Coln St. Aldwyns near Hatherop Castle, where we met the Prince of Wales, Earl Shannon, & the Vale of White Horse Hounds – trotted off to Bibury Race Course Gorse and ran across the country nearly to Jolly's Gorse found and ran into the Bradwell Covers, where I left them & reached Home at 2 p.m. Changed horses, & rode to Heythrop & joined Mr. Brassey & party coursing at 3½ p.m. Dined at Mr. Blake's, & Home again at 1.30.

Bot. 20 qrs. Sainfoin Seed of Mr. Bulford who was Coursing Judge – 19 of us Dined, viz. Blake, 4 Savidges, Sargent, Bulford, Langham, Edgington, Edwardes, Stilgoe, &c.

The Prince of Wales looked jolly and well – and appeared in sound, robust health.

March 1. By 12.20 train from Shipton, to Moreton races, too frosty – did not run – and after Lunching with Sir M. Crofton, retd. and tea'd with Mr. Reade.

March 2. By 12.20 train, and witnessed 4 steeple chases at Moreton in the Marsh, Mr. Goodwin won on Sir Morgan Crofton's 'Duellist' with Lord Beresford on 'Gt. Eastern' 2nd both *first class riders*.

March 7. Wild squalls of rain, sleet, & snow.

The Boynal Gardens piece sown with a cwt. of Nitrate Soda, & 1 cwt. Salt, per acre: on Wheat stubble for Barley.

The Revd. L. Baker, Ramsden: called, on his way to meet the Hounds at Gawcombe.

March 15. Drove to Chippg. Norton and rode to Banbury . . . Bloxom was the largest village on the way, its Church has a fine spire.

Had Champagne with Mr. J.P. Barford – saw Mr. Stilgoe, Durchfield, Eagle, etc.

Dined at the Unicorn, the 1st Dish, *Yorkshire* Pudding – 2nd Dish, Fish, &c – 2s.

Walked round 'Banbury Cross'.

Mr. Gregory, Enstone: gave me a ride to the Mason's Arms – and Mr. Prior hence to Chippg. Norton – bought 4 sorts of Cakes.

March 18. Fine, cold, bright, drying March day.

Drove Alice, John Chas., & Grace, to Charlbury Church, went by Ascott & Chilson, retd. by Forest & Leafield: each in 45 minutes.

Charlbury is a very old, badly built stone village & Quakers are said to abound.

March 26. Met Hounds in Forest: ran up & down & lost near Singe Wood – came hence to Fairspear tried Priest Grove blank, but found a yellow looking Dog Fox in Boynal Cover, ran him thro' Priest Grove, over Langley & South Lawn to Hensgrove, Stockley, back to Hensgrove, Widley, Falsgrove, Asthall Leigh & killed him in Fulbrook after a deal of very good hunting – making 117 killed this season – Mr. Brassey gave me the Brush – some of them Luncheoned as we came past Fairspear to try – been here 3 times & found every time!!

Planted 2 fir trees in Corner Holt.

Had conversation with Mr. A. and H. Brasseys, the Honble. Victor Spencer, & Mrs. Sartoris.

A *dead heat* between Oxford & Cambridge – for the first time on record, on Saturday last.

April 2. Accompanied Alice, Elizh., John Chas., & Grace, to Witney, and witnessed the Athletic Sports – the *10 feet* high jump & Steeple Chase, was 1st class.

Tea'd & 'Moselled' at Mr. Shillingford's.

April 3. A year to-day, since I arrived with Mr. Wm. L. Mason at Fairspear 9 a.m. for Breakfast & Valuation, from London, it appears a *very long* year, to me.

Miss Reade, her only brother Joseph, and a visitor, Tea'd, and spent the Evening with us.

Heavy showers, from 3 to 6 p.m.

April 9. Too wet for Thrashing, or Drilling.

Met Hounds at Tangley, ran a Fox 20 m. and killed him – tried Barrington Park covers blank, but found Biscuits, Bread & Cheese, Sherry, & Ale at the House – I came away for Shipton Sheep Auction of 1519 – Mutton 8 pence per lb., *out* of *Wool* – Hounds went to Rissington, blank, and found at Gawcombe – Hounds ran away from all the 'Field' & killed in 45 m.: making 122 Foxes killed in 121 days Hunting: which excels any previous Heythrop Season!!!

Tea'd at Mr. Maddox, & met Mr. Abraham after his tour with Spurgeon, to France, & Italy.

April 11. Drove to Chipping Norton and attended Board of Guardians & Waywarden Meeting.[3]

April 15. Mr. Josh. Reade, Shipton, 62 years old last Tuesday. Rode over and Tea'd with Mr. Mrs. Savidge, Sarsden. Ugly land to farm in wet seasons.

April 20. Fine still balmy morning – heard the Cuckoo, the Turtle Doves, & Swallows, have returned.

John Chas. climbed, & took 5 eggs out of a Magpie's nest, in Shipton Barrow clump.

Finished drilling all, except 30 acres of last Turnip land in Kingswood.

April 28. By 11.28 train to Oxford.

Dined with Mr. Bulford at the Clarendon, and paid him £79.10.0d. for 20 qrs. Sainfoin Seed, which I drilled with Barley, twice over, in Shipton piece.

Paid Mr. Smith for Yarbro' Print, framing: and he accd. me to Mr. Maslin's the Photographer – where I had my Likeness taken.

April 30. Commenced grubbing up the only remaining Eleven Big Oaks, in Big Field, or Calvertt's Corner.

Mr. Brassey, & his Yeomanry comd. Drilling for 8 days: they practice at Blenheim Park, & stay at Woodstock.

May 1. Fine sunny cold morning.

12 Ploughs in Big Field.

Upwards of 50 Lambs have been worried by Foxes, or Dogs, around here – I have 4 killed.

Tea'd with Mr. Reade, Shipton: and walked all round the Grove, & Gardens, &c.

May 4. The famous 'Jack Morgan' from Grove, and Hamblin from Badminton came by 12.15 train – Morgan showed me his Gold Watch presented to him on 12th April by Earl Scarbro' along with £600, with £200 to £400 more to follow, subscribed by all those who have hunted with him the past 20 years in the Southwold and Grove Counties – he really looks as active and hearty as 20 years ago, when he comd. with the Southwold.

May 5. Morgan left by 8.15 train, and I drove Hamblin, to Heythrop Kennels: Hazleton showed him all the Hounds, & he took the 11.15 train from Chipping Norton for Chippenham.

Dogs are dying at Leafield from Poison, given by owners of Lambs – lately worried to an alarming extent.

May 8. Heard the Turtle Dove – our Summer visitors are all here, & I hope Summer will shortly set in – Turtle Doves, Swallows, Cuckoos, & Nightingales, all migrate in the Autumn.

May 9. On Monday Alice, Eliz., & John Chas., & myself witnessed the Oxfordshire Cavalry Review in Blenheim Park, 183 assembled, Mr. Barnet, Glympton, is Colonel: for which Mr. Gering said he was paid £1000 a year.

May 14. About 2,500 sheep sold at Shipton – Tea'd with Mr. M. Savidge, at Mr. Reade's – Delightful rain in the Evening.

May 19. Mr. R. Morris came, and measured the 7 Oaks taken down in Home Lawn – 364 feet, at 2s. per foot, and I take Bark, Lop & Top, for labour of cutting, lopping, & barking.

Lost several lambs, off Ascott Common Clover and a many are being worried by Dogs, or Foxes, around here, but cannot *see* them destroyed, altho' watched.

May 21. Alice, Eliz, Grace, & John Chas. accd. me, to Over-Norton Park: and we witnessed Athletic Sports, and races, and Tea'd at Mr. Edwards' with Messrs. Gregory, Sargent, Savidges, Blake, &c – 'Merry Bell' was the winner of 3 races.

May 25. John Chas. and I, by 8.15 train to Breakfast with Mr. Oliver, Oxford – who afterwards showed us Keeble College, Christ Church, Radcliffe Library – Great Tom & Tower, Sheldonian Theatre – Addison's Walk – Broad Walk – Magdalen College, Park, Deer, &c – the Reredos in All Souls very elaborate – altogether had a most interesting day, & reached Charlbury at 8.39 p.m. – very gratified with all I had seen.

May 26. The 2nd Club Day at Leafield – 2 Bands played here. . . .

The warmest, finest day, this Spring.

May 30. Lord Falmouth's, Sylvio, won the Derby.

Messrs. Hornsbys, sent Mr. Geo. Black to repair Drill, Reapers, Ploughs, Engine, etc.

Dined at Mr. Reade's Rent audit, Shipton Lodge, with Messrs. Bliss (2), Maddox, Baker, Savage, Iorns, & others.

June 2. The 2 shepherds been clipping all the week.

John Chas. walked round with me in the evening, by Boynal Cover, Priest Grove, and Home Lawn Spinney, I shot a Turtle Dove – for examination & curiosity, & very pretty, quickflying, wild, harmless birds, they are, arriving every Spring with the Nightingale, Cuckoo & Swallow – I also shot 5 Rabs.

The Revd. Baker, Ramsden, called: about the Lamb-worrying, in the District.

Mr. Mrs. T. Brassey reached Hastings last Saturday after a sail round the World, in 46 weeks, & their yacht 'Sunbeam' traversed 36,000 miles.

June 3. At Leafield Church in a.m.

John Chas. accd. me to meet Mr. Bliss at Cornbury Park 3 p.m. – we found Mr. Holland, who shewed us round the Garden, House, &c: and regaled us with Ale!!! Brewed in 1856!!! Tea'd with Mr. Bliss.

June 5. Drove John Chas. via Brize Norton, Clanfield to Farringdon. We walked round the 'Folly' whence we beheld a grand view of the Counties all round – saw the 'White Horse' on Hill side – Dined with Mr. Rd. Eagle at 'Salutation Inn' &c – we drove away by Buscot Park &c – to Northleach – over the 'Half-penny Bridge' and back, to say we had been in Oxon – Berkshire – Wiltshire, and Gloucestershire: during our Drive to this point: passed Lord de Morley's seat, just outside Northleach, & came thro' Filkins, to Burford and reached Home 6½ p.m. – after a *very instructive* day's drive of about 40 miles.

June 9. About 120 acres Turnips Drilled.

Drove Alice to Ranger's Lodge Cricket Ground & saw a match between Merton Coll. Oxford, & Lord Dillon's Club – nice ground.

Fine, mild, sunny, afternoon.

June 12. John Chas. and I, to Ascott & away by 8.15 train to Oxford, Didcot, & Chippenham, were met by Mr. Chas Hamblin, & driven to Badminton Kennels, saw a magnificent pack of Foxhounds, they hunt 5 days a week, & killed 180 Foxes last season – we saw all over Badminton House, Church, Park, Gardens, Worcester Lodge, the Monument Cherry Gardens, &c – the Red Deer are very large – the Hinds were dropping their 'Calves' – in all sorts of places, & nooks in Park. Saw Mrs. Vincent, late Miss Morgan, Grove.

June 13. Rode the Marquis of Worcester's Polo Pony, & John Chas. rode Lord Henry Somerset's Polo Pony. We went to Cherry Gardens, round Park to Worcester Lodge, there is an Avenue 3 miles long, extending from Badminton House to Worcester Lodge, we also read the Monument inscriptions – it was erected to the memory of Lord Henry Somerset, who fought in Peninsular wars & finished all up with Wellington at Waterloo – Lord Raglan was a Duke of Beaufort's son – they trace their pedigree from John of Gaunt – left Chippenham Station at 3.10 for Didcot, Oxford, & Charlbury at 8.39 after a very pleasant visit to fine old Badminton – everything gives place to Foxhunting – no Game – Lord Worcester is in Guards, takes his vacation in the Winter & hunts his Father's Hounds 5 days a week: his only Sister is Marchioness of Waterford.

June 25. Very busy Clover carrying, mowing, & Turnip Drilling – the most critical time for Farmers and the time I always feel the most anxious.

July 2. By 8.15 train to Oxford Assizes – sat on a Horse Case Miles Pestall, v. Long – Lopes, Judge: special case.

July 9. 10 Beast sold in London, £242.18.6d. *net*, & 13 Sheep at Shipton £42.18.6d. *net*.

Mr. Fred. Matthews drove up from Shipton, and farmed me round, before he returned to Taynton.

July 10. Miss Mason, Calceby – Miss M. Fowler[4] – Alice, Elizh., & John Charles accd. me to Mr. Abraham's and heard Mr. Spurgeon, preach to about 3000 persons under the Oaks at Ringwood.

July 12. Alice, Elizh., Miss Mason, & Miss M. Fowler went to Blenheim – they assisted in welcoming home Lady Rosamond, who had been married to Mr. Fellowes in the morning – returned by Charlbury and met Miss Eleanor Gilbert, at 5.40 p.m. from Lincoln.

I paid Tithe to Horne, at Crown, Shipton – Dined: and walked with Mr. Hambidge, & tea'd at Mr. Brooks.

Drove Miss Gilbert round the Farm.

July 17. Mr. Jas. Shillingford came, & weighed 940 fleeces wool 33s. 189½ Tods Wool.

Misses Gilbert, Mason, Fowler, & Alice & Elizh. drove to the Rose Show, Chipping Norton.

July 21. Misses Mason, Fowler, Alice & Elizh. were driven to Ranger's Lodge Cricket Ground, in Whichwood Forest, Ascott, & Worcester Railway employed, were playing a match.

John Chas. & Grace, tea'd at Mr. Waller's.

July 28. Lady Churchill sent me a Fawn – being the first I have had given me, thro' Life.

Very nice growing weather.

Aug 1. Called on Mr. Waller 5 a.m. and he drove me to Abingdon – we hired to Ilsley – average show of sheep, which fetched good prices – strong useful Lambs 40s. to 44s. – Ewes 45s. to 60s.

Mr. Newton, Crowmarsh: let a Down Ram Lamb for £42 & only to be used during *this* month!!!

Was surprised at the great size of the Downs, & Cotswolds.

We tea'd with Mr. Widdows, as we returned Home.

At 6.10 p.m.: House Commons rose, after sitting *26* hours, thro' Irish obstruction.

Aug 4. Saw the first piece of wheat 'Fagged' at Abingdon, as we returned from Ilsley Fair on Wednesday.

Mr. Hollis, Coggs, near Witney: rode over and inspected my Land & Drills – previous to my purchasing his all-purpose, Show Drill, at Banbury next month.

Aug 13. Attended Shipton Auction – mutton 8d. – 9½d. lb.
Tea'd with Mr. Jas. Shillingford, & Aitkin at Mr. Price's – looked at his Horses, Wool, & Garden.
Alice, & Elizh. attended a Pic Nic – on Ranger's Lawn, by invitation – about 70 attended: and enjoyed Dancing, & a very pleasant evening together.

Aug 23. . . . 2 Reapers – and *all* hands Harvesting.
Very fine drying day.
Drove John Chas. & Robt. Wm. to Witney Mkt. Ordered a new 8 feet Drill of Messrs. Hollis, Coggs: for Corn, & Turnips drilling £44 complete.
Had Wine & Melon, with Mr. Jas. Shillingford.
Mr. Mace, Milton: recommended *Cotton Cake* for feeding sheep on Turnips.
The Turks continue to harass, & entirely stop the Russian advance into Turkey!!![5]

Aug 29. Fine sunny breezy day – most desirable for the Harvest, at so critical a time.
Miss M. Oliver accd. Alice, and called upon Mrs. Waller. Master Josh. Reade, dined with us.
The Turks are astonishing the World, with their vigorous onslaughts on the Russians, both in Asia, The Balkans, & Plevna District.

Aug 31. Comd. & carried the 27 acres Farthings Wheat – pt. of Stackyard field – drying wind.
Deal of rain, with Floods, in many Counties.

Sept 1. Thirty years ago, as to-day, I commenced my first shooting season, at late Mr. Richd. Barrett Adams', Collow Grange; whither I had gone to farm for 2 years on the 14th of the previous July.
I shot 5 Hares & 7 Birds – and *to-day*, have been out with my 2 Sons – & killed same number of head, viz. 2 Hares, 5 Birds & 5 Rabbits.
What changes occur in 30 years!!! Mrs. Adams has buried her 2nd Husband, Mr. Robt. Bills, Linwood, the past fortnight. I spent 2 very pleasant happy years with her at Collow Grange – and a nicer man than Mr. Adams, never existed.
Mr. Mrs. John Iorns, Mr. John Clinch, Mr. John Wright dined with us.
65 acres Wheat carried – 550 Wethers on Turnips.

Sept 5. Dragging Turnips – tieing Oats in p.m.
Reaper commenced again at 9 a.m. in Oats by Leafield.
Drove John Chas. & reached Banbury at 8 a.m. – the Harvest quite as backward all the way as at Fairspear – Lunch with Mr. Perry, & called at Mr. Barford's.
The Oxfordshire Agl. Show: very good show: saw Lord Willoughby de

Broke – the jumping was cleverly performed – £1070 in Prizes – altogether the largest meeting ever held previously as a County Show.

Sept 10. Fine cool dull Harvest day – Reapering & *all* hands tieing.
Spent ½ hour at Shipton Sale.
John Chas. shot 4 Rabs. & a Pigeon.
The Turks continue to *harass*, and *check*, the advance of the Russians: the sacrifice of Money & Life, will be enormous on both sides – and *all* for *very little gain*, to *either* party!!!

Sept 12. Three Oxford Down Rams to 230 Ewes – never so early. Drying day with rain at 5 p.m. . . .

Sept 26. Carried 'Harvest Home'. . . .

Sept 27. Rode & reached Heythrop at 10 a.m. Bot. a Blk. Horse of Mr. Blake.
At Witney in the afternoon.
Commenced thrashing the Barley I carried yesterday to prevent the Clover heating it.

Oct 4. Fair and Mop at Witney – hired 2 Labourers. Dined at the Marlbro' with Mr. Blake.
Lambs & feeding [ewes] comd. in Calvertt's Corner Turnips.

Oct 13. A nice rain last night, which will nearly suit everything & everybody. All work well forward for winter.

Oct 15. . . . A terrific West wind, last night, and early this morning, wrecked the corn ricks, &c: all around.
Mr. Baughan hauling the last 2 large Oaks off Home Lawn.
From 1st Septr. to this date, killed 219 Rabbits, 83 Partridges, 24 Hares, & 11 Pheasants, besides woodpigeons.
Ground too hard for wheat drilling!!!

Oct 20. Met Vale of White Horse Hounds at Alvescott at 9 a.m., called at Mr. Nalder's – found 6 or 8 Foxes in the Furze Cover – left them running at 10½ and with my men drilling wheat in Hall's Hill, at *12 o'clock*.
By 1.24 train to Oxford – Bot. Hunting Hat of Juggins. Game a/c recd. £8.15.4d. of Turrill.
Had a bottle of 1863 Port, with Mr. Blake.
Beautifully fine, *sunny, mild, still* day.

Oct 23. Very showery – most suitable weather for the land, & cropping – Turnips growing.
Isabel, vaccinated at Leafield!!!

Oct 29. Wind & rain *all* the day – could not attend the 'Opening Meet' at Heythrop – they have been out 17 times, & killed 25 Foxes, the shortest cubbing season ever experienced – hindered by the *late* Harvest, & then too *dry*, for both Hounds, & Horses.

Oct 30. Mr. Blake, Heythrop called & bot. 2 of my *old* Horses valued from Heanley – recd. £25. . . .

Suffering my *first, violent* attack of Lumbago.

Nov 6. Finished Wheat Drilling – Cockshott Hill – Hall's Hill – & Boynal & Brazil – 103, 95 and 61 – 259 acres.

Dined at Mr. Reade's, Shipton Lodge – with Messrs. Goodwin, 2 Blisses, Maddox, Baker, & Price & Secker –

Slight showers, which prevented my going to Mitford Bridge – hunting – to meet Lord Willoughby and his Hounds, & call upon Mr. Harbidge, Little Compton.

Nov 13. Met Lord Willoughby & his Hounds at Weston House – found Fox in Rollright Spinnies, & killed him in road at top side of Whichford Wood – had some talk with Lord Willoughby de Broke – found a 2nd Fox in Whichford Covers, ran round & up thro' Green Hills & away for Braile's Covers – I left and returned by Little Rollright, and saw the *Old* Druidical Stones, on Mr. Bliss's Farm: they are famous Relics of past ages: came thro' Over & Chipping Norton – altogether enjoyed the day – the weather was quite sunny, and hot.

Nov 15. Drove to Witney with Elizh. – Pony ran away for 1½ mile beyond Minster nearly to Witney, 2 black pigs and then, the noise of a 4 wheel, lent me by Mr. Looker, further frightened him – eventually pulled him up.

Judged at Mr. Hemmings, Witney, Root show: Mr. H. gives the prizes to his customers who purchase Lawes' manure of him.

Dull, *hazy, misty,* mild day – !!!

Nov 19. Mr. Price inspected Turnips & Sheep, & tea'd.

Shot 4 Rabs. in Home Lawn Spinney, & one out of chamber window, by *moon*light – a new plan adopted with all rabbits, which stray into shrubbery as they feed on the *front lawn at night.*

Nov 24. John Chas. and I, by 8.13 train to Reading at 9.37: inspected a splendid show of Roots at Messrs. Sutton's Establishment – Retd. to Oxford 2 hours, & reached Ascott at 5 p.m.

Wind & rain all day. . . .

Dec 11. Jane, Eliz., & Self left Ascott 8.13 train, & accd. Mr. Savidge & Bryan to Paddington Stan., went to the Salisbury Hotel, and reached Agl.

Hall Cattle Show at 12 p.m.: Saw several Lincolnshirites.
Witnessed 'Sardanapulus' at the Duke's Theatre.

Dec 12. With Mr. Blake to Windsor Fatstock Sale on Her Majesty's 'Show Farm'. Found the Queen & Princess Beatrice examining the Beast & Sheep for Sale, when we reached the yard – she looked hale, & well – saw the Dairy, Aviary, Cows – &c.

Luncheoned at Mr. Tait's – &c – took the under-ground rail from Paddington Stan. to Farringdon St.

At St. James' Hall in p.m. – heard Mr. Simms Reeves sing 3 times – Santley – Mme. Sterling – &c.

Dec 13. Elizh. Anne, and I, walked into the City before breakfast, and viewed St. Paul's – Mansion House, Bank, Exchange, Post Office &c.

Left Ludgate Hill Stan. & spent the day at Crystal Palace, witnessed 'Our Boys' for the 939th time at 3 p.m. In the evening we saw 'Pink Domino' performed at the Criterion Theatre.

Dec 14. From Farringdon Stan. to Aldgate St. for the Tower, and afterwards walked to the Old Swan Pier, and took packet for Westminster Bridge – 'Bus by the 7 Dials to Oxford St. Bazaar, & back to Westminster Aquarium. Zazel shot from the Canon etc.

Witnessed Mme. Angot in the evening.

Dec 15. Mr. John Wright, Haugh: Linc: accd. us Home. Did some shopping in a.m. – reached Ascott at 5 p.m.

Dec 29. Rain during the night, & thaw this morning.

By 1.24 train to Oxford. Saw Mr. Jonas Paxton & arranged to forward a Truck of Fat Sheep every fortnight to his Oxford auction.

Dec 31. John Chas. & I, drove to Mr. John Bryan's, Southleigh. I walked round Farming with Mr. Bryan and his brother James from Shilton – and John Chas. accompanied Mr. John Bryan jnr. round the Farm & Tar Wood, with their Guns – but only shot, brace Hares, & Rabbit.

Talk – Rubber at Whist – and left at 9.30 after a very pleasant visit. . . .

1877 was wet and cold, in the Spring – a hot, short, Summer – very wet *latter* Harvest – wet all through November, & rain, snow, & wind last 8 or 9 days of December – bad yielding Harvest, thro' *frosts* in *flowering* time – very few Apples, scarcely any Stone Fruits, no acorns – but abundance of Nuts, only!!!

1878

Jan 1. Mild dull day – like November.

Rode to Charlbury Stan. & Fair – formerly famous for its show of Horses – a few Horses, 3 Beast, & about 50 Sheep were all I saw at 11½ a.m.

Jan 7. Rode & called on Mr. Blake at Heythrop who accd. me to meet the Hounds at Pomfret Castle – found in Cold Harbour Gorse, & ran to Hook Norton, Swerford, Tew, & lost at Deddington – came away Home, & called at Mr. Ward's, Chadlington.

Accd. Alice & Eliz. to Witney Childrens Ball, reached Home at 2.20 a.m.

Jan 9. Called, & Mastr. J. Reade, Shipton: accd. me to meet the Hounds at Bledington Mill, tried Icomb cover blank – found at Gawcombe, & ran around Ripington nearly to Clapton Gorse lost – but found him again in the Gorse, ran a mile and a half, & killed him, dead beat, in a Turnip field – tried Gawcombe again, Wick Hill covers, & Botany Bay, blank –

Had some talk with Mr. Henry Brassey – Hall – &c – Mr. H. Brassey's first day with us, he comes and resides at Copse Hill for 8 or 10 weeks on purpose to hunt with his Brother, & visit him at Heythrop Park –

Mild day with storm clouds – wind N.W.

Jan 26. Wind & snow last evening, turned mild during the night, a slight frost, & very little snow left this morning – *Nothing* so *changeable*, as the *weather*.

Farming around all day: the Dores, placing a thorn hedge across Kingswood, at 2s. per chain.

Jan 29. Started 8.30 a.m. & reached Mr. Harbidge's Little Compton 10.15 – rode away together at 11 – and met Lord Willoughby & his Hounds in Wolford Wood – ran a Fox to the Village & round the wood until we killed him – found 2nd Fox in a cover a mile off the wood – ran thro' the whole wood & away at South end – turned about, & ran to Moreton-in-Marsh – & I left them running & rode on to Batsford, examined & admired Lord Redesdale's House, Park, Church & village, called at Mr. Schofield's – while Hunting *rode round* 4 Shire Stone – which marks the boundary of Warwickshire, Worcestershire, Oxon & Gloucestershire – Returned Home thro' Moreton, Stow-on-the-Wold, Gawcombe, Fifield, & Milton – Home at 4.15 – after a very various pleasant day's ride, in nice, *bright, still* weather.

The view is magnificent – from Batsford House, & Park top.

Feb 2. A slow thaw – no snow, or rain –

Sold 300 fat sheep at Oxford & London, during Jany. 4 sixtys on Saturday evenings to London and 2 thirtys to Mr. Jonas Paxton's auction at Oxford: averaged 70s.9d. *all round, into Bank* – 240 of them Wethers, & 60 Ewes.

Miss Gudger, Burton-on-Trent, visiting Elizh. Anne.

Feb 10. Very full of cold – very starved yesterday – kept the House *all* day.

Feb 16. Met Hounds, at Mr. Mason's, Eynsham Hall, the House is magnificently furnished – saw the most elegant Billiard Room I ever looked over. . . .

Feb 19. Called at Mr. Reade's, & accd. him, with Mr. Warton & Scovell, & joined upwards of 70 Conservatives at a Dinner, Crown Inn, Shipton – very pleasant evening.

Feb 21. Drove John Chas. & Robt. Wm. to Moreton-in-Marsh, Steeplechases – a good afternoon's sport – had Champagne with Mace, Reade & Brassey, & Claret with Sir Morgan Crofton.

Feb 25. About 160 ewes lambed, and all doing well.

Sold £7.5.0d. worth, of old wood, to Hill & Co.

Mr. Widdows sent 100 very useful Tegs, making 250 recd. for Summer clovers: I paid £676 for them

March 4. Earl Ducie's only Son, Lord Moreton, 21 to-day. The Tenants recd. £10 per cent, on their last ½ year's rent, as a Souvenir of the event[1]. . . .

March 8. Attended a Leafield School meeting – to audit the a/c's for Government Inspector.

Mr. Wm. Abraham in the evening.

March 9. Farmed all round – and by 12.15 train from Charlbury to Worcester at 1.45 – met Mr. Baughan at the Junction Chipping Norton.

Saw Cheltenham & the Malvern Hills – the place is only 7 miles from Worcester.

Evesham is a good, rich, sheltered soil, & famous for early vegetables, & Orchards, £8 to £12 per acre rent is given – the Duke de Aumale has an estate near here.

The interior of Worcester Cathedral is gorgeous and imposing – the most costly pulpit, I ever saw – and the edifice stands on the banks of the Severn. Saw Salmon in the market which had been caught in the morning – Worcester is a finely situated Town, and said to contain 41,000 inhabitants.

Left at 4.10 and reached Shipton at 8.12 after a very instructive afternoon's travel.

March 12. Mr. Richardson-Gardner's Hounds ran a Stag from New Barn straight to Shipton, Ascott, Pudlicote, & Chadlington: a distance of 18 miles – Horses & Hounds were all beat out.

Jane and I tea'd at Mr. Mrs. Sargent's, Chilson – met Mr. Freeman of Short Hampton – who had been to New Zealand, Australia, America &c – and most interesting descriptions he gave of the climates – manners – &c of the strange medley of mankind, with whom he came into contact during his Travels.

March 13. Sat $5\frac{1}{4}$ hours, on various parish business, in Chipping Norton Union.

Tea'd with Messrs. Craddock, Sargent, and Harris afterwards, at the Crown Inn.

March 23. My Foreman, Johnson Patchett, leaves for 'Reeds Farm', Minting, nr. Horncastle, Lincolnshire: on Monday, after spending upwards of 2 years on Fairspear.

Mrs. Waller & Clark, called.

John Chas. had his last day's Hunting – met the Hounds at Blenheim Pillar – ran Brace Foxes, but killed neither.

March 30. Snow squalls during the day with a high North wind.

Sent 60 Wethers to London last Saturday – 60 to Oxford on Tuesday, 60 sheep to Shipton on Friday, and 60 Wethers to London, to-day.

A great walking match has taken place at the Agl. Hall for a Belt and £500 – began 1 a.m. on Mon. morning and by 8.20 on Sat. evening – O'Leary $520\frac{1}{4}$ miles – Mon. 117, Tues. 83, Wed. 88, Thurs. 86, Fri. 83, Sat $63\frac{1}{4}$ – the second called Vaughan completed exactly 500 miles, and the third called Brown, 477 miles: three great feats of endurance.

Dores £2 for plashing $16\frac{3}{4}$ chains hedge in Big Brazil, next Mr. Hambidge – and 9d. score for tieing Faggots. 705 Faggots at 2s.6d. score & 18s.8d. for Carriage, £4.8.$1\frac{1}{2}$d. and Dores £3.6.3d. for Plashing, & tieing.

March 31. Walked to Mr. Blisses at Waterman's Lodge – he sent me Home in a snow storm.

April 5. John Chas., & I, had our first day with the Old Berkshire Hounds – met at Curbridge – Earl Craven, his Brother, 2 Sons, 2 Duffields, & about 80 others were out – admired Mrs. Strickland's House &c, at Cokethorpe – saw a deal of bad land &c – Treadwell, the Huntsman, is one of the best sportsmen in England.

April 9. Rode over to Sarsden Lodge, & saw Mr. M. Savidge's Steam Plough & Harrow at work. . . .

April 17. Since 30th last Augt: paid Mr. D. Alder, Shipton: £429.2.6d. for

20¾ Tons of Pure Linseed Cakes from Mr. Foster's, Gloucester and 30½ Tons of Cotton Cakes from Hull: at £6.12.6d. & the Linseed £11.5s. to £11.10s.

April 18. Rode from 4.45 to 7.15 a.m. thro' the Forest, Capps Bridge, by Dean Grove to Spelsbury, Taston, Enstone, Tews, Newington & Bloxham to Banbury – breakfast at White Lion, with Mr. Brazier, Winslow: &c – spent my time amongst the Beast & Sheep – saw Messrs. Eagle, Widdows, Stilgoes, Perry, &c.

Rode round, & admired 'Banbury Cross'. Banbury is a very Business Town, for Agriculture, & 10 miles around it, is a district of very productive land.

Returned from Banbury, in exactly the same time as I rode there the other route, 2½ hours.

April 26. Drove Mr. Jas. Shillingford, via Shipton, to Mr. G. Garne's, Churchill Heath, and attended his Short-horn Sale. Earl Fitzhardinge, Lord Morton, Baker, Byass, Barnett and a large company attended – we called at Sir Morgan Crofton's, Shipton Court on our return, & inspected his Horses, articles of Vertu, &c.

April 30. Rode over, & tea'd & spent a pleasant evening with Mr. M. Matthews, Fifield – he has a nice Rookery – and water-cress bed in his Garden.

May 3. Walked down to Mr. Price's – & met Mr. Baker after his Grass Keeping Sale at Shipton – Mr. Savidge bought most of it – for his son Matthew, who enters on the Farm next Michaelmas – Mr. Baker has retired to Cheltenham.

May 15. Very showery: becoming alarming for both Wheats and Fallows.

May 17. Labourers all retd. Home – Manure &c – all worked up – only Spring Corn weeding in hand – the first time I have sent them back, since I've been here – owing to frequent, and continued rains.[2]

May 18. Terrible rioting at Blackburn, Accrington, Burnley, &c, thro' Strikes, and Lock-outs – which I hope will end in a friendly and business-like settlement of the affair, between Capital and Labour. . . .

May 19. Isabel resolved to walk at last!!! and very safely. Fine brisk windy morning, after a very rainy week. . . .

May 23. Heavy rain again last night – doing a deal of injury to White Corn all over the Country – cannot even plough the Fallows.

At Witney in afternoon – bot. a Grindstone of Mr. Marriott, called a *Bilston* 34s.

May 29. Royal Oak Day.

26 Coaches & Fours, had their opening meet in Hyde park, the Duke of Beaufort is President.

Capt. Berkeley died at 4 a.m. in Cornbury House. He had a Lease of Forest, Park, & other shootings, belonging Lord Churchill.

May 31. The finest day – for a long time – May has been a terribly wet, hindering month to Agriculture. . . .

June 1. Dull cold morning – cold blighty wind.

Sheep all washed at Minster – on Mon, Wed & today – turned Bull out onto Kingstanding Grass: and Mr. Eagle sent me 11 Beast for Langley: and Tegs gleaning the Charlock out of Fallows

June 6. At Witney, Sq. Harcourt M.P. addressed his constituents at the Marlbro'.

Paid Mr. Eagle £569.5.0d. & £16.16.8d. for 21 Beast, and 78 Tegs 55s.4d.each. . . .

June 11. Jane and I by 8.15 train to Oxford – engaged a Miss H.J. Jackson – as Governess to Grace and Fred – comes on 1st August.

Spent 3 or 4 hours in Bath & West England Agl. Show. Frequent heavy showers with high wind, spoiled all our pleasure.

I dined with Geo. and Albert Gresswells at Christ Church College – about 180 to 200 sat down – the most magnificent College Dining Hall in Oxford.

Jane went to Mr. Mrs. Oliver's: we ret'd. at 8.17.

June 12. Had 3 to 5 weeks of wind & frequent rains, day & night – to the great injury of Farmers.

Earl Beaconsfield and Marquis Salisbury, gone to the Congress at Berlin – the *crowning act* of the *Premier's political* life.[3]

June 14. Drove Alice to Blenheim Palace and attended a Bazaar: the proceeds of which go towards the Restoration of Woodstock Church.

We saw the Duchess of Marlbro', and four of her daughters, Lady Beaumont, and Fellowes, & Lady Georgina and Sarah, also the Marchioness of Blandford, &c – small attendance – Oxford Show – Money scarce!!!

June 15. Dull cool dry day. . . .

Leafield in High Feast – 3 Clubs & Music!!!

June 19. Eleven men finished 131 acres of Clover and Sainfoin mowing: 3s.6d. Clovers, 3s. Sainfoin per acre – they cut the 46½ acres of Shipton field in 3 days.

June 27. Josh. Clement entered upon Patchett's place as Man of all work – or working Foreman.

June 29. Comd. Clover carrying on 21st ult. and finished to-day at noon – 131 acres, out of Shipton Field, Kingswood, 6 Roods, Little Brazil, & the Farthings – Men 3s.3d. – no Ale.

Spent the evening at Mr. Reade's, Shipton: and he related his journey, and described the present magnificent exhibition now being held on the Champs de Mars, Paris.

Showers with Thunder in afternoon.

June 30. . . . This has been an unusually wet month – the first 19 days heavy rains & showers, then 7 or 8 hot sunny days *here*, with Thunder or Storms all round.

The continued rains of May and June have sadly injured the cropping on clay lands.

July 4. Drove Alice yesterday evening to Mr. Abraham's, Minster, and heard the famous Mr. C.H. Spurgeon's Father preach to about 2000 – his Text was 'Consider' – his Son was laid up with Gout: a hale hearty jolly man of 70 years. . . .

July 17. Been sunny & hot for nearly a month: injuring the late sown Spring Corn, & wasting the pastures, but suiting the Wheats, which are now only worth 38s. to 40s. per qr.

Aug 2. . . . Attended Mr. Brassey's Ram Sale, bot. no. 40 £7.7.0d.

Miss Jackson, Oxford: came as Governess to Grace and Frederic.

Aug 5. Drove Eliz. and Miss Slack to a Garden Party in the Grammar School Avenue, Witney: all went on nicely till 9.30 when a heavy rain set in – when all adjourned their Music & Dancing to the school-room. The rain abated at Midnight, when we left for Home – after being hospitably entertained by Mr. Heels.

Aug 8. Mr. Hambidge tea'd, & lent Ralph & Reaper for 2 days.

Drove Miss Slack & Eliz. to Witney, and Mr. John Clinch showed us over Mr. Early's Blanket and Rug Manufactory.

Aug 10. Sufficient rain to stop reaping & tieing. Weather most unsettled, but suits the late Turnips.

Aug 20. . . . Finished reapering for the present – Fern Hills Barley, and 50 to 60 acres Oats, only left – not ripe. . . .

Mr. Hunt, Fawler, drowned himself – one of the Duke of Marlbro's tenants: on Fri. or Sat.

Aug 22. Commenced carrying wheat with 8 waggons out of Cockshot Hill – 3 Hoers at 5s. per acre began to single Turnips in Boynal Gardens – and 2 skerries going in Witley Hill. . . .

Aug 27. Had a fine, drying *second* day's Wheat carrying.
Some dozen of the Leafield School children came to Tea. Mrs. Hambidge & Daughter, & Sister, drove down to Tea.
Robt. Wm. ill – Diarrhœa & sore throat.
Sir Morgan Crofton, taken a House at Ascott!!!

Sept 7. Fine dry week – cutting & carrying every day – 100 acres Oats, & 60 acres Barley left out.
A Thames Steamer ran down, laden with a Pleasure Party, of upwards of 800 – 700 drowned – awful – the sad calamity occurred opposite Woolwich about 7½ p.m.

Sept 10. 700 persons drowned in the Thames, thro' the 'Princess Alice' Saloon Steamboat, being cut in two by the 'Bywell Castle' – an awful calamity!!!

Sept 12. . . . Advertised in the Oxford Times, for a Shepherd, Head Carter, & Groom.

Sept 17. Took up 214¼ Tods Wool 33s. – Mr. Nash weighed.
Dined with the Bailiffs and about 40 other gentlemen at their Venison Feast, Marlbro' Hotel, Witney.
Yesterday hired a Groom (Wm. Reed) a Shepherd, Wm. Hanks, & a Head Carter, David Eeles.

Sept 25. Rain & showers – nearly all hands at Burford Fair – I stayed at Home. . . .

Oct 3. Carried Harvest Home – by finishing up Barley rakings out of Fernhills – a long lingering Harvest – began on 3rd August – carried all in *good, dry* condition!!! . . .

Oct 4. Commenced thrashing for 10 or 12 days. . . .
Drove Jane, Alice & Elizh. to Shipton Court – Mr. Mrs. Reade shewed us all round the fine *old* large antique mural mansion, built by James 1st as an appanage to Wychwood Forest

Oct 5. Drove Jane to Burford Mop, for 3 hours. . . .
Been a fine, mild week – with intervals of *hot* sun.

Oct 8. Drove Josh. Clement to Faringdon Mop, we hired J. Constable for Cowman, & 3 under Carters. Dined at 6 p.m. – with 30 others, at a Venison Feast, at 'Red Horse' Shipton.

Oct 12. . . . F. Shayler called, & arranged to watch Priest Grove, & Boynal Cover.

Oct 13. Rode over to F. Shayler's, Cornbury: & found they had captured Edw. Pratley & 'Puggy' Moss, both Leafield men – a third escaped. Their 2 guns will be given to me after their prosecution next Wed. Robt. Caldicott was 3rd thief!!!

Oct 16. Philip Moss & Edw. Pratley, convicted at Chipping Norton for Poaching at Priest Grove – 1 month's imprisonment & sureties for 1 year – or 6 months in prison, & brot. their Guns Home – they will also be fined for carrying Guns without a Licence.

Jane hired Mary Ann Bryan for Housemaid. . . .

Oct 24. Very rainy morning – did not go to Stow. Spent the evening at Mr. Reade's, Shipton, and saw Mr. Samuda, the new Tenant at the Court.

Admired some Indian Books given years ago, by Warren Hastings, to Sir John Reade.

Oct 31. Miss Susie Wilkinson came as Governess from Welby nr. Grantham – ground covered with snow around Huntingdon.

At Witney – *Best* dry Red Wheats 40s. top price, good sound average malting Barley 38s., Oats 40 lbs. per Bushel 21s., *black* ones.

October has been a wet, mild, growing month, suited Turnips & young Clovers, better than finishing Harvest & Wheat sowing – ice made its appearance, the two last mornings.

Nov 3. At Leafield Church in a.m.

Black, & Rime frost – fine, cool, clear morning.

Great depression in Commerce, Trade, & Farming, all over England.

Nov 13. Met Hounds at Mr. Foxe's, Bradwell Grove. . . .

Became acquainted with Mr. C. Samuda, Shipton Court on my return home – Mr. Samuda, Price, Sir Morgan Crofton & myself, called & had Sherry at Mr. Cheetle's, Burford.

Mr. Reade walked down in the evening – yesterday he had shaken hands with Prince Leopold at a Mason's Lodge meeting, in Town.

Nov 16. Paid Labourers 2s. per day for the past week – being 1s. per week lower than ever before, since I came here.

Forwarded 20 Green Rounds, for exhibition at Messrs. Sutton's Root Show on 23rd inst.

Nov 17. Drove over & dined with Mr. Richd. Eagle at Stanton Harcourt, ascended Alexr. Pope's Study – looked over the Kitchen, & the Harcourt Aisle in Church, a most Historical place – the Harcourts trace their pedigree back for 35 generations.

Nov 23. By 8.13 train – reached Reading 9.45 – took 1st Prize at Messrs. Sutton's Root Show, for their Imperial Green Rounds – brot. away a Claret Jug. Left at 11 for Oxford Market. Very slow depressed Corn Market – walked round the Meat Market with Mr. Savidge.

When I reached Home filled the Glass & Gold Claret Jug with·Champagne, and Drank Miss Susie Wilkinson's health on her 21st anniversary, also my brother's 47th.

Nov 30. Rode over to Shipton & met Mr. Ellis, bot. 2 Colts. Mr. Robt. Hambidge, Ascott: rode down & looked at sheep, tea'd, and stayed the evening – we arranged to petition the Crown for a return of percentage on our Rents.

November has been famous for *heavy* rains & snow in many parts of England.

Mr. Gye, accidentally shot at Dytchley, he is famous as a Theatre Manager, at 'Her Majesty's'.

Dec 7. . . . Col. White Melville, famous for Hunting & writing thereon – broke his neck while hunting with the Vale Hounds the other day – his Horse suddenly pitched him on his head, while galloping over a ploughed field.

Dec 11. John Chas. and I – went by rail to Oxford – attended the Christmas Auction – sold 60 Wethers 66s.4½d. Had Luncheon with Mr. Squire at Wadham College. . . .

Dec 12. John Chas. to Witney – left £430.8.4d. at Bank. Had hard black frost since Sun. – thick air, and ¼ inch of snow.

Horses hauling Faggots, Thorns & Manure. . . .

Dec 24. Severe frost, with snow, all over England, and extra so, in Scotland.

Yesterday, Albert Brassey Esq. sent me a Brace of his well-fed Heythrop pheasants.

Dec 25. Alice, Grace, Jane, & 3 Sons, & Self, at Church. Wind, snow, sleet, & rain in the afternoon.

Dec 31. Met Mr. Brassey and his Hounds at Bledington. . . .

1878 has been one of the worst on record for Commerce, Trade, & Agriculture. . . .

1879

Jan 2. Miss M.H. Fowler,[1] Alice & Eliz. accd. me to Witney.
Sold 20 Theaves £62.11.6d. & bot. 6 Pigs £5.2.0d. Dined at the Marlboro', met with a Druid, Mr. Blake, Bliss, Nalder, Empson, etc.

Jan 4. Miss Fowler went to Witney in search of her lost Watch & chain, & rings – cried it[2] – had a dozen bills printed at Mr. Wright's – when she returned to Fairspear, saw it amongst the snow, as she stepped out of the waggonette, close by the Front Door. . . .

Jan 29. Drove over to Chipping Norton – returned back to Sarsden Lodge, & tea'd with Mr. M. Savidge, who I found laid up of a bad cold. Report says Mr. A. Brassey has given £30,000 for a large Yacht: the 'Czarina'.

Jan 31. Only ploughed 5 days the past 8 weeks.
Rode down to Ascott Stan. and on to Chilson, tea'd and spent the evening with Mr. C. Sargent, met Mrs. Huckvale, & John & Kate Sargent.
Frost and Snow nearly all the month.

Feb 7. Miserable weather – Rain – Fog – rained & then wind.
Took the Chair at Conservative Association Dinner, Shipton, about 56 attended – Sir Morgan Crofton Bart., & Revd. York, Jo. Scovell, Plowman, & Spackman, etc: enjoyed an hour at Mr. Reade's afterwards.

Feb 12. Wrote for 'small seeds' for Witley Hill.
Wm. Wright, Wold Newton, Linc: died of concussion of Brain – from a fall from his Horse, ae. 46.
Met Hounds Bradwell Grove, found Fox in big wood, and ran round till we killed – found 2nd in Jolly's & ran to ground in East Leach – we found & chopped 3rd in Jolly's – & ran a 4th from Jolly's, & after a deal of running, lost in Williamstrip – had Whisky & Biscuits with Dr. Cheetle.
Mr. Price walked up in the evening.

Feb 17. Mild still morning – an *inch* of *snow* fell at 6 a.m. Have 40 to 50 lambs already.

Feb 19. Alice & Eliz. accd. me to a Ball in Witney Corn Exchange. 121 attended, & all passed off most satisfactory – reached Home 7.45 a.m. – 2 inches snow.

Sheep Market, Witney, date unknown, but pre-1904. *(Oxfordshire County Libraries/T.E. Worley)*

Feb 21. Shot a hare and 6 rabs. in Priest Grove.

King Cetewayo of Zulu, & his soldiers, attacked our Camp, slew 500 men – took away 1200 rifles, 2 cannon – lots of ammunition – 1400 beast – 100 waggons, £60,000 worth of Stores, &c – the other of our forces had all left in search – & missed their enemy: large reinforcements are being sent, and a *sound thrashing* is in store for those semi-barbarians.[3]

March 6. At Witney – Tea'd at Mr. Abraham's.

72 Officers and 906 English soldiers were massacred at Isamunda, in Zulu, South Africa – we shall eventually make a terrible example of them, but I fear it will cost this country a terrible expence.[4]

March 11. Witnessed the School inspection at Leafield by Mr. Picard, & Ely – & lunched with Revd. Lee.

March 15. Mr. Price accd. John Chas. & myself, to meet Hounds at Ditchley Park – ran & killed brace Foxes in cover behind the Hall – ran a 3rd from Oaklands & lost, tried Glympton Gorse blank, & left them at 4 p.m.

Mr. Price called, and had Moselle, this being his 56th birthday – a nice, agreeable man, & neighbour.

March 18. The Honble. Victor Spencer, only Son of Lady Churchill, Cornbury: officiated as Page at the Royal Wedding last Thurs. – fulfilling which office, entitles him to a commission in the Guards.[5]

On Sun. John Chas.' Ferret escaped out of his Box, and worried the 2 Tame Rabbits.

March 21. Rode with Mr. Price to Farmington Grove – ran a Fox to Aldsworth and lost – found again in Clapton Gorse & left them running for Lodge Park.

Sherried at Mr. Huckvale's – and remarked a Tree in front of his House on which Bruern parish papers are always exhibited.

The 3 Mr. & 2 Mrs. Brasseys out.

March 22. Wychwood Parish meeting held at my House 7 p.m. Messrs. Waller, Wm. & Benj. Hobbis, & W. Ellis attended.

Dry with cold North wind.

March 24. Trimmed out the Shrubbery previous to Mr. Clutton's visit – he wrote to say an allowance of £200 would be allowed me, by the Crown, for the year ending Michaelmas 1878, as a recompence for the bad seasons.[6]

March 28. Alice, Eliz., Grace, & Fred, went to Sanger's Circus, now visiting Witney.

March 31. Shipton Auction, 1237 sheep – 5s. per 8 lbs.

Cornbury House, south front from the park *c.* 1894, taken by Henry Taunt *(Oxfordshire County Libraries)*

The past month has been seasonable, reasonable weather.

April 7. Mr. Pegler, W. Bliss, & Handcock, are all quitting their land at Shipton, Milton, & Churchill – all owing to the wretchedly depressed times, & *Bad* seasons.

April 8. The Zulus continue to make raids amongst our men, they slily attacked a convoy of waggons etc. & killed the whole party – over 100 men, &c.[7]

April 12. Snow began to fall at 7 a.m. and continued *all* the day – frost at night – had to put all the Beast in Crews, & Ewes & Lambs, in Shelter.

April 13. *All* the ground covered with snow.
Snow in Paris – a most backward, hindering season after one of the severest winters on record.

April 15. *Extraordinary* morning – 3 inches snow on the level – such a fall of snow, so late in Spring, has not been seen, during my lifetime, of nearly 50 years.

April 16. Mr. Price walked down, & tea'd.
The Zulus have killed 70 men & 7 officers, of Col. Wood's relief convoy: but am expecting to hear of a reckoning day, now that our reinforcements have safely arrived.[8]

April 18. Began only this morning to drill the Big Field Oat stubbles, with Barley!!!
Met Mr. Brassey and his hounds at Longborough Gate to finish the season – killed brace Foxes, from Crawthorns tried Seizincote blank – found at Evenlode Main & ran to Cornwall & killed – found, ran, & lost, from Addlestrop Hill – gave up 6½ p.m. bid Mr. Brassey good bye, & Mr. Price & I accd. Mr. Byass to Dalesford House & had Sherry etc. – reached Home at 7.50 – been on my Horse 9 hours & 40 min!!!!

April 21. Lord Chelmsford, & reinforcements: relieved Col. Pearson, at Ekowe – the Zulus made a grand attack upon our Forces, when within 7 miles of Ekowe – but were badly beaten back, with great loss – on *2nd ult.*

April 27. Rode thro' the Forest, to Shayler's, the Keeper's; saw a Wild Duck on its nest – Lady Churchill reached Home last night, from her visit with the Queen, to Baveno.[9]
Cornbury Park was formerly a part of Wychwood Forest when its extent was 12 by 9 miles – the Mansion was built by Earl Danby on the site of Cornbury Hall where in 1588 died Dudley Earl of Leicester, the husband of

Amy Robsart, & a great Courtier of Queen Elizabeth. In 1622 the House became the property of Earl Clarendon whose daughter was the mother of Queen Anne; and in 1751 the Estate was purchased by the 3rd Duke of Marlborough – George the 4th Duke, left it to his youngest son, who was created Baron Churchill in 1815 – his eldest son, the present Francis Geo. Spencer was born in 1802: and in 1849 married Lady Jane Conyngham – their only Son Victor Albert Francis Charles, was born 23rd Oct. *1864*!!

The spacious stone entrance is adorned with armour and Trophies, of the 1st and great Duke of Marlboro'. A highly decorated, richly wood-carved Chapel joins the House, the Park is full of Deer, and in parts charmingly picturesque.

April 28. Sadly pained with rheumatics – can neither sit, stand, or lay, comfortably!!!

Seen *one* Swallow, but not *heard*, the Nightingale, Cuckoo, or Turtle Dove. . . .

April 29. Mr. S. Haigh called, and gave me advice, & Medicine, for Sciatic Rheumatics – constant pain in right leg, side, and back.

Cold drying day – wind North – like early *March*.

April 30. Sadly afflicted with rheumatics in my right side & leg, in Bed till 11 a.m., & in House nearly all the day.

April has been cold, frosty, foggy, snowy, rained & been most irregular throughout – everything is backward, and the present farming prospects very disheartening.

May 1. Frequent snow storms from 4 to 10 a.m., with cold North wind – 45 acres Barley, & 85 acres Oats, to Drill!!!

The Woodman Dores, cutting & tieing all the Thorns and refuse, in Home Lawn spinney, at 1s. score.

May 6. Joseph Arch came from Oxford, & addressed the Leafielders.[10] Cold drying North wind. . . .

May 21. . . . John Chas. drove Alice, Eliz., Grace, Miss Wilkinson, and myself – to Blenheim and we attended the Agl. Show held in the Park – the Duchess of Marlborough and her Daughters joined us for hours – several of the County Gentry attended – 400 sat down to Lunch; A.W. Hall M.P. is quite an Orator – Mr. Price showed Alice round – my Rheumatics pained me very much – walked in the Gardens with Miss Wilkinson, & Grace – reached Home at 7.20.

May 26. Cetewayo sues for Peace – hope it may prove true, and put an end to a *very expensive*, but *useless* war.

May 28. Mr. Reade's Rent Audit – too unwell to attend, still suffer from Rheumatic Gout, on right side, from Hip to end of little toe.[11]

Very growing & showery.

May 30. Rode to Shipton Auction.

Mr. Reade's Tenants very disappointed at rent audit, no return being made.[12]

Mr. Waller rode over, and described the merits of his Farm to me.

June 16. . . . Rain for hours – can neither plough nor work the land, weed corn – clip sheep. The Country in the worst state I ever knew it in my life, seasons are ruining English Farmers, & Farming.

June 19. This is only the 2nd fine dry day, for some time – Large mkt. at Witney – Wool 22s. to 24s. – Mutton 9d. lb. . . .

June 20. 42 years this day, since the commencement of Her Majesty Queen Victoria's reign.

Heard the sad news from Mr. C. Sargent, that Prince Louis Napoleon, had been assegaied, by Zulus while reconnoitring with Bettington's Horse – in South Africa – so dies the last Napoleon – and a Nation's Hopes and Destinies[13]. . . .

June 21. Drizly, cold, and windy, like March. Drove to Burford – for an hour in p.m.

Only finished sheep-shearing to-day!!!!

June 24. In the evening the groom drove Alice, Eliz. & Grace to the Forest Plain – the Hawthorns are in full bloom, we met with Shayler & Keepers, & upwards of 800 Pheasants being reared on the Plain – we walked to the Seven Dials, rides, or lites.[14]

June 30. John Chas. & myself left Ascott at 8.13 and reached Royal Show, Kilburn 11 a.m. Saw the Prince and Princess of Wales, & 3 Daughters.

Witnessed 'Venice' at the Alhambra – scenery gorgeous.

Stayed night at Salisbury Hotel, Fleet St.

July 1. . . . By train from Blackfriars Bridge to Queen's Park, West Kilburn – attended Members Meeting at 2 p.m. the Prince of Wales, President: he spoke capitally.[15]

John Chas. and I, took train at Kensal Green for Broad Street & walked on to the Tower, & Ridgeways – bot. specs – 'New Babylon' amused us, at the Duke's Theatre.

July 2. We joined Mr. Clutton & son George,[16] & about 18 others at Charing Cross for Redhill, hence over Earlswood Common to Mr.

Clutton's land at Reigate &c – Luncheoned at his Brother Henry's, & retd. with Mr. Tom Caswell & met train at Redhill for Town – shopped &c – & left Paddington Stan. at 6.30 for Charlbury at 8.39.

Very showery day – the Show yard in wretched condition – most unfortunate for all concerned.

July 9. Rode over to Cornbury Park, & joined F. Shayler & Keepers, Fawn shooting – killed 16 – I shot at 7 and killed 6 – the first time I ever witnessed the sport – rather resembles *lamb shooting*!!! . . .

July 11. . . . Comd. Drilling Turnips, by Forest, in Cockshott hill – Comd. turning Sainfoin in Shipton piece.

Jane, Miss Bettinson, Wilkinson, Eliz., Grace, John Chas., & myself, met Mr. Price, at Cornbury House 5 p.m. The Housemaid showed us all over the ancient edifice, which is one of the most historical Homes of *Old* England – it contains paintings, armour, & a great variety of *knick-knacks*: we retd. partly thro' the majestic Beech Avenue

July 18. Gave 3s.6d. for 46½ acres Sainfoin, & 3s. for 93½ Clover mowing – 12 scythes – £22.3.0d.[17]

July 20. Heavy rain in the early morning – and terrible showers during the day – the most awful season – for Frosts, cold, & rain, during April, May, June, & till present date – ever experienced by the present generation.

July 24. Forwarded rent to Whitehall – £200 returned for the year ending *last Michaelmas.* At Witney.

July 25. Lord Chelmsford attacked & slaughtered 12 to 1500 Zulus, burned Cetewayo's Kraals at Ulundi, & driven them into the Bush – which will virtually end the tedious war.[18]

July 26. Finished Turnip Drilling, 115 acres, & fallowing 48 acres owing to the continued rains from *April* till *now*, never experienced, indeed the past 4 months have been the most hindering for Agriculture on record. . . .

July 31. Showers during the night, & rain *all* the morning, cleared up at noon.

Joined Mr. Mrs. & Miss Margaret Abraham: and we drove thro' Bampton Buckland, & Standford to his Son George's at Sparsholt – tea'd & stayed night.

Admired the Wilderness – Spring, called 'Black Jack', and secluded woody village – looked over part of the Farm.

Aug 1. Up at 5.40 & off through Wantage to Ilsley – large show sheep – slow trade – very good rams – lambs 30 to 35s. each & very good ones –

cheaper than I have seen them for *years*: returned back thro' Wantage, & admired the Statue of King Alfred – cost £2000 – Population 3,800.

Dined at Mr. G. Abraham's, Sparsholt, & we all reached Ringwood at 10 p.m.

Sir C. Throckmorton's seat at Buckland is well kept and has much wood – & the finest Cedars, & Wellingtonias I ever saw – Home at 10.30 p.m.

Aug 6. Jane & I, by 8.13 train to Oxford, Didcot, & Chippenham, and on to Badminton Agl. & Flower Show – Morgan & Treadwell there, after Judging the Puppies the previous day. Stayed night at Mr. Hamblin's.

Aug 7. Rode round the Park with the new Entry – 8 of them broke away after a red Buck – &c: saw a Fawn shot: saw all over the House, Chapel, Stables, Gardens, Kennels, & twice met with the Marquis of Worcester. Badminton is famous for its Hounds, which hunt 5 days per week – the Marquis being Huntsman.

Saw 2 carriages slipped at Chippenham, when the train was going over 60 miles the hour – felt a curious sensation – as if our 2 carriages were running away.

We left *Grand* Old Badminton at 1.45 – after a very pleasant visit & good look round.

Aug 8. Finished up all Sainfoin & Clover carrying, after *40 days*, from commencement of cutting. . . .

Aug 13. At Oxford Ram sales – slow trade for 2nd class sheep. Mr. Gillett, Cote: showed the best Oxford Downs, & had by far the best sale – 42 averaged £13.14.0d. each.

Lightning cut down a chimney, & did other damage at Mr. Clinch's House, Witney – Mr. East's and Dutton's were also injured: 5 persons knocked down.

Aug 22. Continues the most cursed, ruinous weather, on record – cannot thrash Oats – plough Fallows – skerry Turnips, nor even carry manure on the Clover land for Wheat!!! – and this state of things been going on *all over* the *Country* since *last April*!!!

Aug 26. Attended Bampton Fair – large show Horses and very slow trade.

Aug 31. Fine, cold, breezy, sunny morning – At Church.

A terrible hindering ugly month.

Sept 8. Mr. Eagle met Mr. Hardinge here, and they bot. 28 Steers & Bullocks – to my great relief – as keeping is, & will continue, scarce.[19]

Sept 10. Commenced cutting Barley (scythes) with all hands in Witley Hill – the only day this week – showers and rains frequently prevailing – to the great loss and ruin of Farmers in general all over England, Ireland & Scotland – the accounts we read in the Papers are really alarming.

Sept 13. Only cut Harvest *one* day this week!!! – to-day worse than ever – ruin to all Farmers – such a rainy Spring & Summer, has never occurred in my time – and so little sun – for weeks and months together – Sales of farms all over the District, & County. Groom, W. Reed: Cowman, Constable & Labourer Hunt leave at Michaelmas – also Williams – Jas. Dores returns to Leafield, & Ph. Pratley drops from High Lodge, to Kingstanding.

Sept 16. Carried Little Brazil Wheat, for seed.

The worst Harvest I ever gathered – neither quality, or yield, & acres of the wet heavy veins, *worthless*.

Sept 20. Horses 4 days this week ploughing Seedland 2nd time in Kingswood – Hands cutting *all* the week, 3 fine, & 3 misty days.

Sir Morgan Crofton Bart., called: while driving round with Mr. Price – gave him a Landrail.

Very few young Partridges – all ought to be spared for breeding – except now & then, a taste for Selves: the worst season for *everything* and *everybody*, during my *lifetime*.

Sept 25. At Burford Fair – Hired Saml. Hicks as Carter – gave 3 married Carters 13s. week, 50s. for Harvest, 20s. for Ale, & House & Garden. John Chas. & I went with Mr. Gillett, Tangley: & dined at the Lamb Inn.[20]

Sept 27. Mr. Serle – Mather, & Ward, resign their Farms at Chadlington, under Earl Ducie – all tired!!! Farmed for years, and *realised nothing*!!!

A Parson nr. Bampton, has resigned his 'Living', because unable to Farm, or let, his Glebe.

Sept 30. Attended late Mr. Hunt's Sale, Fawler: his famous Oxford Down Ewes averaged 63s. – ram lambs 42s., an eighty guinea [ram] made £7.10.0d. – a great *sacrifice* for *all parties selling* off.

Oct 1. At Witney in a.m. Reduced my Fire Insurance £1,850. Tea'd with Mr. Reade, Shipton Lodge.

Oct 4. Carried Mayes Close Wheat – making only the *3rd Wheat Rick* to *this date* – and 100 acres of Wheat, Oats, & Barley to cut!!! Hope I may never see the like again. . . .

Oct 15. Attended Mr. Hancox Sale, Sarsden: good quality well grown Cotswold Ewes, tupped, *40* to *45s.* each, not seen Sheep, & Beast, sold so cheaply for years.

Tea'd with Mr. Abrahams, at Mr. Savidge's.

Oct 20. Too wet to carry corn.

To Oxford by 1.7 train – bot. Rheumatic clothes at Mr. Key's Holywell Street.

Oct 23. Alice and I, 2 hours at Witney.

Mr. C. Sargent, Mr. Hambidge & Major Bicknell, spent the evening with us – the Major told us a deal about the Zulus. He escaped without a scratch – a very intelligent officer.

Oct 27. Matthew Dores, finished reaping the Colne Wheat in Leafield Piece – sown last Autumn on Turnip land.

Finished Oat carrying out of Ascott Common also carried Oats out of Boynal Gardens, and Boynall Hill – had 9 Waggons, and 4 Carts, at work. . . .

Nov 3. Rode to the opening Meet at Heythrop: saw Earl Redesdale, and Lord and Lady Clanmorris – and all the Nobs of our Hunt – Alice & Eliz., were driven – and looked round the interior of Heythrop House, and enjoyed their visit at the 'Meet' very much. I reached Home at 1.20 and saw the *last load* of the Harvest come into the Yard and much relieved I feel – Thrashing, Thatching, and Wheat Drilling – goes well – *ground Dry*!!!

Nov 8. Rode hence to Swinbrook Mill in 37 m: from there to Mr. Nalder's Alvescot in 37 m also – and met Mr. Hoare & Vale Hounds – they hunt 5 days per week – 10 Horses out every day: found in Gorse & ran, & killed at Clanfield – Kelmscot blank – found in Lord D'Mauley's Park, & after ½ an hour in, & out, the Farringdon covers – I left amidst shouting, & views in all directions, came Home in 1¾ hour, thro' Filkins, & Burford.

Mr. Price in evening – who had had a wretched bad day's sport, with the Heythrop, at Cogges.

Nov 15. Actually had *3 weeks* fine weather – Frosty, & can scarcely plough. . . .

Nov 19. Met Hounds at Mr. John Bryan's, South Leigh: found in Tar Wood – ran to ground at Stanton Harcourt, a terrier bolted the Fox, and we lost in village of Northmoor – found & killed next Fox, in a cover in South Leigh & ran another hence to Cogges Wood, thro' Mr. Mason's covers around Eynsham back to Cogges to South Leigh & round into Witney, out & away to New Mills on the Windrush, where he was drowned.

Nov 20. Snow fell ½ an hour – the first of the season.

Gardener, Thos. Dores, ill of Bronchitis.

Nov 21. Snowy morning – scarcely any Frost.

Isabel, 3 years old – walks and talks well – says 'Luty' for Miss Wilkinson – 'Lack' for Grace – 'Lallah' for Alice, & 'Wupper' for Supper.

Nov 30. . . . Walked up and tea'd with Mr. & Mrs. Reade at the Lodge, Shipton. Had about 18 average days weather during this month – which all things considered, has been the least hindering, of the previous 6 months!!!

Dec 1. Very *severe* frosts – no snow – skating general.

Thrashing – carrying 4 ricks Barley out of Leafield Piece, to Kingstanding, filled the covered shed with Barley straw, and the rest ricked in large Crew.

Manured Farthings for Beans.

Dec 4. Mr. Sherwood told me on Mon. last – that he had not consigned *one* Truck of grain away from Ascott Stn. since Harvest!!! such an event unknown since the rails were laid.

Attended our Annual preliminary political meeting held at Churchill Arms, Ascott, on Mon. last.

Dec 8. Attended Shipton Auction – several rotten Ewes sold at 18s. to 25s. each!!! what a sacrifice for their owners.[21]

Tea'd with Mr. Price.

Dec 13. Took Gun round Farm & accidentally shot couple rabs. in Woeful Green, a Hare in Kingstanding Lawn, & a rabbit out of Home Lawn Thorn Bush.

A fortnight to-day since Horses could plough, hard as ever, no snow, nor wind: the last 5 days thrashed 4 Barley Ricks out of Big Field.

Dec 16. Rode round Farm, & along the Forest valley to Cornbury Park, & Shayler's: back by the Hall to Ranger's Lodge, Waterman's Lodge, & called at Mr. R. Morris' of Smallstones Farm, & on to Churchill Arms, Ascott, where Lord Churchill's tenants were paying Rent, 5 per cent returned!!!

Dec 20. By 11.28 train to Oxford Xmas Market – *splendid* show of Meat, & Game, &c. Bot. a 12½ lb. Turkey of Turrill, 15s. and a pair of Boots at Elliott's, for 9s.6d.

Robt. Wm. joined me at 1 p.m. from Windsor School.

£80,000 at Magdalen College, & £40,000 at St. John's, is being expended, in re-building portions of these grand old places of learning.

Three weeks to-day since ploughs were stopped by this continued hard Frost, with scarcely any snow, and no rain, nor wind!!!

Dec 28. Mr. Hockley and I, rode to Mr. Price's, and Mrs. Hockley, Alice & Elizh. attended Church – after which they wined with us, at Mr. Price's,

Shipton.

At 7.10 p.m. the entire Train from Edinboro', while crossing the Tay Bridge, was carried by 13 spans of Girders to the bottom – every Carriage, girder, &c: left no vestige behind of its whereabouts – about 100 persons are supposed to have perished – not *one* alive to report the awful catastrophe – the most *appalling* railway accident of this *most wretched* year.

Dec 31. Wind and rain most of the day – no work done!!!

So ends the most *ruinously* ugly *seasoned* year, of *this century*.

1880

Jan 2. Met Mr. Brassey and Hounds at Tangley – found Fox in Taynton Quarries – ran him 15 m, and lost him again in the Quarries, whence for ages, stone has been dug & dressed for building purposes – it is said a deal of stone was taken from Taynton, to London, for St. Paul's. . . .

Jan 7. Met Hounds Ledwell village – found 5 Foxes in Worton Heath – ran one to ground near Lower Worton Church – tried all North Aston blank – & all the Covers including Wilson's Gorse – the northernmost Cover of the Heythrop Hunt – found in a Gorse field at Adderbury, ran about a mile to ground – the country along the Cherwell valley looked in a *ruinous state* – many of the sheep are rotten – a most terrible year for the Farmers. Called & had Pork chops & Sloe wine with Sir M. Crofton at Mr. Geo. Dawkins, Enstone: reached Home 7.20.

Jan 17. Continues favourable weather for Farming.

Great many rotten sheep all over the Country, coupled with bad quality & yield, especially of Wheat & Barley, causes Great Depression in Agriculture.

Jan 31. Could not plough in Decr. and this makes 11 days delay *this month* – keen, foggy, to clear, no rain, sleet, or snow.

Thrashed last 3 Barley Ricks out of Big Field, carried them to Machine in yard, to keep the Horses exercised. . . .

The Country is in a wretched state of *Depression*, caused by the past 3 or 4 adverse Seasons – to finish: Beast, & young Horses & Sheep, that have depastured very *wet*, or *flooded* lands, are dying of the *Rot*.

Feb 1. Fine sunny day, with continuance of hard frost. Branches broke off the Oak, Elms, Larch, & Fir trees: thro' the frozen Fog ice, collected on the stems, &c. Never saw the sight before – no wind – the weight was too much for the frozen boughs.

Mr. Price tea'd, & gave report of Sir M. Crofton's Horse Sale, at Chipping Norton, last Fri.

Midnight £84 Surbiton £46 At Last £45 Glenalmond £54 etc.

Feb 7. . . . The Honble. Victor Churchill officiated as Page at opening of Parliament – he told me he danced all the previous night at Heythrop Fancy Dress Ball – off by 1st train on Thursday to London, & after opening Parlt. returned to Cornbury Park to Dine – see what Railways enable us to perform, in 12 hours!!! . . .

Feb 9. Mr. Mitchell came from Birmingham, and joined us for a Ball, at White Hart Hotel, Moreton in Marsh, I went through the invitation of Mr. John Scovell – with Alice & Elizh. – left Ascott 5 p.m. train, and we retd. by 7.35 train next morning to Shipton – a very pleasant agreeable party of 62.

At Shipton Auction – sold 18 cull ewes 47s.6d. net.

Feb 21. . . . Tinson, a soldier retd. from Zululand, came in, and described the attack at Ghingilova & Ulundi, &c: he escaped without a scratch – he knew Lord Chelmsford, Genl. Crealock, Colonels Buller, Pearson, & Wood, and Capt. Kevil Davis, &c. An armed Zulu carries 12 assegai on his back, a dagger in a case, a Shield & 6 assegai in his hands – they are a tall, stout, active race.

Feb 29. This is the 5th Sunday in Feby: which will never occur again, in my lifetime!!!

Feby. has been mostly mild – with heavy rainfalls, all over the Country.

At Leafield Church in a.m.

March 9. Rode by Cornbury Park, Dytchley & Steeple Barton, to Mr. Darthwaite's, Rousham: met Mr. J.R. Young, Ailby, nr. Alford, Lincs – and looked over Mr. T. Berridge's farm, at Middle Aston, 514 acres: about half arable – & Grass & Meadow. Recommended Mr. Y. to bid £800 Rent, & 3s. per acre will cover other expences: 10 miles from Banbury, & 12 miles from Oxford – Clement U.C. Dormer is owner, & a very pleasant man I found him – returned from Hopcroft's Holt by Ludwell Farm, to Glympton, & Ditchley, &c.

March 12. Robt. Wm. accd. me to Mr. Gillett's Sale of Sheep, Horses, & Implements: some of the best specimens of Cotswolds, & Oxford Downs I've yet seen: Ewes & forward lambs 58s. to 65s. – prices very various as a doubt existed as to their being *free* of the *Rot*!!!

March 25. Good Friday: men, & horses, worked in a.m. Fine bleaching north east wind, with sunshine. Plenty of dust.

March 31. Rain at 11 a.m. 200 acres Spring Corn in, *March* has been very favourable for the land.

April 2. Started 10 a.m. in wind & rain, to meet Hounds at Longboro' Gate – left wet coat at Adlestrop Station – cleared up a fine day – killed 1st Fox in 25 m. at Seizincote – ran a 2nd to Trafalgar Farm, Guiting: and left them running 3rd from Crawthorne's for Banks fee – Home at 6.45 – Had Sherry &c at Mr. Goodman's, Banks fee.

Attended a Conservative Committee meeting, at Crown Inn, Shipton 7.30 – present Messrs. Brooks, Brown, Haigh, Reade, Dee, Waller, Lardner, Morris, &c.[1]

April 5. John Chas. to Heythrop, saw the 80th Fox killed. Mr. Brassey broke his Horse's neck by falling over a wall – the Huntsman and 1st Whip, have each killed a Horse, during the Season.

April 24. Very fine week for Thrashing – Rolling Wheats and stirring Fallows, & Farming in general.

Lord Hartington summoned by the Queen, to form a Liberal Cabinet!!! The Duke of Marlborough leaves Dublin Castle next Wednesday for London & Blenheim Palace – we shall be glad to have him in Oxfordshire.[2]

Thos. & Jas Dores, settled up for cutting *last* 2a. 3r. 28p. at south east end of Priest Grove: they are paid 40s. per acre for cutting down – 6 pence per score for smooth and thorn faggots – 6 pence per hundred for thatch sprays – 3 pence per score for sorting out the small poles, 1 penny per dozen for Hedgestakes & Hurdle stakes, and a half penny per Bundle for Peasticks.

The 2a. 3r. 28p. contained the following: 3480 smooth Fags: 2235 Thorn Faggots – 5360 Sprays – 101½ Dozen Hedgestakes – 60 small Poles – 48½ Dozen Hurdle stakes, and 33 Bundles of Peasticks – also sold £5 of Ash. . . .

April 28. Continues most favourable for cleaning the land.

The result of the General Election is 353 Liberals, 62 Home Rulers, and 237 Conservative – 652.

May 4. Mr. A.W. Hall will oppose Sir Wm. Harcourt on his re-election as Home Secretary – Harcourt foolishly introduced his friend Chitty, & lost Mr. Hall his seat by only 9 votes a month ago – so Hall has now resolved to fight, & oust Harcourt!!!

May 5. Mr. Lardner poisoned by accidentally drinking a glass of Ale, mixed accidentally by the servant with Petroleum – at Chipping Norton.

May 8. John Chas. & Robt. Wm. accd. me to Oxford, in Pony Trap, and witnessed the contested Election between Harcourt and Hall; the Polling was conducted with great diligence and good feeling – to the delight of the noisy multitude, the result at 7.10 p.m. – from the Town Hall window, was Hall 2735, Harcourt 2681. . . .

Cold dull dusty day – reached Home at 9.30 – after a very Excitable day's amusement.[3]

May 10. Shipton Auction – Mr. Neill, Fox, & Shillingford, called with me, at Mr. Price's.

Mr. A.W. Hall, went up this morning to Town and took Oath & Seat, for Oxford City – amidst great cheering – when he reached Oxford in the Evening, the multitude dragged his Carriage Home!!!!

May 11. Heard the Turtle Dove in Boynal Cover, for the first time this Spring.

The Nightingale sings *daily*, & nightly, most brilliantly in the shrubbery – altho' the weather continues *cold*.

May 29. Royal Oak Day . . . Jane and 3 children to Witney. John Chas. and Robt. Wm. joined the Cricketers at Leafield – Leafield beat Bampton – Mr. Worleys 2, & Headley, drove down here for the evening, afterwards.

May 31. Been very dry throughout this month, with cold winds, & occasional fits of sunshine.

A heavy rain, or showery week, much required all over the Country.

June 5. Deal of cold wind, & slight showers during the week – heavy rains have fallen in *many* districts of the Country.

Wheats and Beans cleaned: 50 acres Swedes and Turnips drilled: the remaining Fallows dragged &c. – Beast all out at grass – and nearly all the sheep clipped – work forward.

June 15. Mr. W. Abraham lent me his Charlock Machine. It cut 3 parts of the Charlock – and with a little *more speed*, and *rather sharper sections*, would be *perfection*. . . .

Messrs. Hambidge, Gillett, Young, Brookes, Price, Waller, Gomm – & others, came to see the Charlock Machine at work. . . .

June 18. The 65th Waterloo anniversary: what *Deaths* and changes, & Wars – since that brilliant Victory.

Fine breezy growing day – trees rapidly approaching full foliage. Only finished Clipping yesterday!!

Mr. William Abraham called, to see after his Charlock cutter.

June 29. Rode round in the Evening, and walked round Mr. Waller's Farm – wonderful Season for Charlock, and Thistles.

June 30. Mr. Gladstone's Ministry rapidly declining in the Public's estimation – Hares & Rabbits Bill withdrawn – Malt Duty probably will follow, and the Bradlaugh Trouble[4] will prove the climax – the Budget will never be passed from present appearances!!!!

June has been a moderately cold, fine month, with sadly too little rain around here – altho' Storms and Floods have occurred in some Districts.

July 1. At Witney Market.

The Carbridge Estate of 884 acres with Houses, Cottages, & Buildings – bid £20,000 – reserve £42,000!!!! that shows the state of the Country!! and the Depreciation in Land!!!

July 2. Mr. Richd. Eagle junr. drove Mr. Bateman Senr. here and bought 14 steers.

Yesterday morning finished Clovers mowing for the 1st time over – in Witley, & half the Big Field, 2s.6d. per acre.

Turnips in Mayes Close, & Little Brazil, doing well: half Greystone, & half Green Rounds – for Lambs. . . .

July 5. Met Mr. Brown, Burford: to revise the Voters' lists for several adjoining Parishes, at the Crown, Shipton – Messrs. Reade, Waller, 2 Hambidges, Gillett, Savidge, Brookes, Young, Franklin, &c: joined us.

July 7. Miss Alderton, Finstock – brought an Action for Breach of Promise, against Mr. Chas. Hunt, Fawler: £400 damages allowed!!!

July 10. Weather *showery*, & very *drying* – carried Clover out of Big Field, 6 hours in p.m.

4s. per acre for Turnips singling – 2s.6d. for Clovers mowing – 2s. for daily labour.

A Devonshire Parson does not pray for our present High Court of Parliament – because, he says, they are such a moderate lot!!!-?

July 27. Shipton Odd Fellows' Fete – Drove Grace & Fredc. down to 'Short Croft' for an hour – shot at a Trio of Pipes, & broke one.

Racing – Music – Comic Singing – Wire Walking – Negro Minstrels – Zulu Donkeys – &c.

July 30. Eliz. drove to Charlbury with Miss Bullock to meet her brother Mowbray – ran on Entrance Pillar when returning into Yard, & broke gig shafts.

Mr. Clutton wrote and fixed 12th August for Rent audit – 20 per cent to be deducted off the next year's Rent due next Michaelmas, & payable in Jany. 1881.

July 31. . . . July has been famous for storms, & frequent rains all over the Country – been most disastrous to the Clover and Hay crops, and is now telling most unfavourably on the Harvest.

Great heat prevails in New York – several have daily died of sunstroke – here we scarcely have a sunny day – this country has had no regularity of seasons, this past 5 years!!!

Aug 4. Finished thrashing Oats, and my 4th and *worst* Harvest at Fairspear.

Some nice coveys of Partridges – appear strong & healthy: Grouse are said to be *healthy, strong, & numerous.*

Aug 14. Turnips *all* skerried, 20a. to single in Hall's Hill.

80 acres Clover down – 25a. Ascott Common Fallows twice dragged – 40a. Shipton Sainfoin ploughed – with *fine weather*, can work *all* up in 4 days: & begin Kingswood Wheat. At Burford.

Aug 16. The Hares & Rabbits Bill will pass[5] – to my great surprise: and I hope the Malt Tax Repeal, will benefit all.

Aug 19. . . . Carried 2nd crop Clover out of Big Field 77a. – better crop and better quality than the 1st cut. 2s.6d. per acre – 8 men cut it in 6 days, each time, at same price. . . .

Aug 25. 2 Reapers finished Kingswood 110a. Wheat – in 3¾ days, and commenced Farthings Oats: 5s. per acre, for tieing and raking all up.[6]

Aug 28. 6 Waggons, 12 Horses, 20 men, 4 boys, & 2 women – commenced carrying Kingswood Wheat.

Mr. Price rode round Farm, in the evening: looked at the Lambs and Feeding Ewes on Turnips for the 1st night – also at 450 breeding ewes in Leafield piece (Clover), and examined cropping. Mr. Price lost 3 games Bezique, with me!!!

Sept 6. Fine weather – all hands cutting & carrying, 7 waggons: most fortunate for the Country being blessed with such a magnificent opportunity, for gathering the Harvest – which taken on the whole will be better than it looked – or we expected!!!

Sept 15. Spent 1½ hours at Chipping Norton Market – Red Wheats 38s. – very *good* Barley 33s. – rather miserable prospect!!!

Sept 16. John Chas. & Robt. Wm. accd. me to Mrs. Freeman's Sale, Shorthampton – which is one of the best farms owned by Lord Churchill.

We looked over the Church – one of the 3 Bells had been broken over a Wedding Peal – in even that secluded, dilapidated place.

Sept 18. No Harvest work done *this week* – Barley sprouting, stacks wet – prospect bad – many parts of the Country in high Flood!!!

Sept 23. Carried 16 loads Barley & Oats – the first since *10th ult.*

Sept 25. Attended Burford Fair for 1½ hours and hired a Carter, called John Willett: all yearly men less money.

Carried Boynal 30 acres Oats yesterday, & to-day.

Finished cutting White Corn, & Beans – the whole of Cockshot Hill Oats & Barley – & 20 acres Beans to carry.

Alice returned from Lincolnshire on Thursday – brought very bad tidings – of Floods and spoiled crops.

Mr. Price won 4 games Bezique 2s.!!!

Oct 4. Shot with Mr. Price over Shipton: Messrs Reades and John Rowland joined us – dull, glass falling, Birds extra wild before change of weather – only shot 8 Birds and brace Hares!!! saw a nice show Partridges.
Dined at Mr. Price's.
Carried 2 ricks Oats – only the refuse left out, and 20a. Beans – rain at night.

Oct 6. John Chas. & Robt. Wm. drove to Witney – hazy day. *All uncarried corn*, spoiling in the fields – rain, haze, Fog – never Dry!!!

Oct 15. A fine still dull day – carried Farthings Beans, and finished Barley rakings – and *glad indeed* to say – carried 'Harvest Home' – after having bad hindering weather, ever since the 11th September.

Nov 1. At 9 a.m. rode away for Ranger's Lodge, Dean Grove, Pratt's Bottom, Lidstone, to Heythrop Hall: rode down the Park, and examined the newly-built Church and Rectory, back to House and down the Enstone 'Drive' which is the prettiest in the Park – altogether Heythrop is the *most complete 'Country House'*. Killed 3 Foxes by 3 p.m. – when I left.
Rather amazed at Influenza, preventing my Horse from being put in the Stable.

Nov 3. Mr. T. Eve, wrote me a doleful letter of Floods, Wrecks, & Ruin in Farming in Lincolnshire – these seasons & Foreign Competition will sadly affect *English* Agriculture.

Nov 15. I weigh 10 st. 5 lbs. – Jane 11 st. 12 lbs. John Chas 9 st. 6 lbs. Robt. Wm. 8 st. 7 lbs. Grace 6 st. 6 lbs. Fredc. 3 st. 4 lbs. Isobel 2 st. 6 lbs. S. Wilkinson 9 st. 10 lbs.
High wind with heavy rain at 3 p.m.
Ireland I hope will be able to alleviate her *opression* by *Agents* and *Landlords* – who are daily proved to be the *Root* of *Discontent*!!!

Nov 16. Rode and met Lord & Lady Willoughby de Broke and Hounds at Wolford – ran round the Wood & Village and lost at Dunston Coppice – found again at Aston Hale, ran out for Batsford &c – lost in high wind – they went on for Golden Cross, and I came away and reached Home at 4.30 in a high cold N.W. wind.
Floods all over the Country.
Lord Willoughby has 2 Sons and 4 Daughters, and 1 Brother, and 4 Sisters: he is the *Best* rider, and gentleman Sportsman I almost ever beheld: but very arrogant in his style in the Field.

Nov 30. Dined with 600 Conservatives at Woodstock – Lord Salisbury, Jersey, Gibson, & Chaplin were the *Best* Speakers – altogether a most successful meeting.

Dec 9. Ireland in a state of Anarchy & Rebellion. Life and Property – unprotected in some parts: Law & Order at an end in others!!! . . .

Dec 14. Drizly rain in a.m. – heavier in p.m. – but mild.
John Chas., Grace, Fred., Eliz., and Miss Wilkinson attended a Concert at Leafield!! *native talent*!!!

Dec 15. Attended Board Guardians, Chipg. Norton: after which appealed against my Shipton land – only found 56.2.24 – instead of 60.1.1 – the Assessment Committee altered the *acreage*, & reduced the *Rateable* Value.

Dec 22. Wretched misty, *cold*, rainy morning, from the East. Received a letter from Gilbert Mackinder, New Zealand: written on 4th Novr. Stated his Gooseberries were ready for pulling – Oats 1s. to 1s.6d. per Bushel – the exporters were doing well with them in *our market*, altho' 16,000 miles off!!!

Dec 31. Brilliant sunrise, on 2 inches snow, with very slight frost. The past year has been famous for Floods – and we have had 5 years experience of them, & find they are far more injurious to farming than extra Hot, or Dry seasons!!! Such Depression in Agriculture, never existed since I was born – we do sincerely hope this new Decade may bring more propitious Times, or the Farmers, & Farming, of Old England will be irretrievably ruined. Farms to let in most parts of England – a deal of land daily going out of cultivation.

It is years since this Country had a heavy Fruit Crop – Wheats range from 38s. to 46s. – Barley 24s. to 36s. – Wool 28s. – Beef 9s. to 10s.6d. stone – Mutton 8d. to 9d. lb. – Labourers 1s.10d. to 2s.[7]

1881

Jan 15. Thrashing Barley, and Manuring young Clovers, *all* the *week* – *hard frost,* with an *inch* of *snow.* . . .

Jan 17. To-day and Saturday – the 2 coldest days during the past 20 years – and lowest Temperature – as taken in Hyde Park – Jany. 15th – 20 Degrees – 16th 16 Degrees and 17th 20 Degrees again!!! . . .

Jan 18. A cold East, North East, cutting wind, set in at Midnight, and continued blowing with *Great Violence* from Noon to Midnight, with snow again!!! . . .

Jan 19. . . . Yesterday, the *most violent* on record here, & to-day *much* the *same,* with *less* wind, and snow nearly the whole day – Thatch off Ricks – Tiles – Slates – and much damage done.

Jan 20. Could not ride or drive, to Witney, or Ascott – for snow: no trains to London on G.W. for 3 days – no letters reached here from Monday to Wednesday at noon!!!

Wrecks on the Coasts all round England – and Damage on Land is terrible to relate.

Jan 22. Frost more severe every night – Snow digging from Parish to Parish. Water pipes, all frozen, in Home Yard.

Jan 24. Great loss of life, on the Coasts, all around England, as well as in many parts of the Country – 27 persons are said to have perished on Salisbury Plains – besides *hundreds* of sheep.

Jan 27. Drove thro' the snow, with difficulty, to Witney: dined with Mr. Wm. Worley – no Auction – all markets prohibited for a Time – to check spread of Foot and Mouth Disease.

A *slow* thaw set in – *Ground itself very Frozen.*

Jan 31. Not been able to plough, since the 7th ult, for Frost and snow!!! but begin again tomorrow – this has been one of the most *severe* Januarys, on *record*!!!

Feb 1. Tea'd with Mr. Reade, Shipton Lodge: and accd. him to a Conservative Comtee. Meeting, at the Crown Inn – arranged matters for our Society Dining at Ascott on Feby. 15th.

Feb 3. At Witney – regretted very much, to hear of Mr. Willm. Shillingford, having failed in the Wool Trade!!!

Feb 5. Met Hounds at Shipton Barrow, found and ran thro' Larch Holt – away to Widley, thro' Hensgrove and Stockley, & lost in Ast[h]a[l]l: found 2nd Fox in Fallsgrove, ran to Hensgrove and Stockley – after a deal of running – killed south end of Stockley. Mr. Dawkins, Mather, Huntsman, &c – called as they passed – for Home.

Been an 'Extraordinary' week in Parliament – Members sat from *Monday* to 9.30 on *Wednesday* morning – or 41 *hours*.

The *whole* of the *Irish* Members – 36 – were suspended on Thursday, amid such *scenes* as were *never witnessed* in the *House* of *Commons*.[1]

Mr. Edwd. Askew Sothern, died on 20th January, was born on 1st April 1830 – famous for the original performance, of 'Lord Dundreary'. I admired him perform it – several years ago!!!

Snow nearly all melted.

Feb 6. Still silently suspicious day – Could hear Ascott Clock Bell, sure sign of *change* – sooner or later – or 3 to 24 hours.

Feb 7. Wind rose at 6 a.m., a Hurricane by 10, with *driving* snow, all men off the land by 11 a.m.

Feb 14. St. Valentine, or Candlemas Day.

Wind & rain, *all* day – wonder if we shall have a good Apple year!!! Wind on this day is said to be a precursor of Apples. . . .

Feb 15. . . . Attended Shipton Conservative Assocn. dinner, held at the Churchill Arms, Ascott: J. Scovell, in chair, about 67 sat down – very convivial evening – present J. Reade Esq., Haigh, Forshaw, J. Brown, Waller, Morris, Revd. Yorke, Compton, T. Brooks, Hambidges, Hobbis, Cole, Stratton, &c.

Jane spent evening with Mrs. Hambidge, Ascott, Mrs. Forshaw there – and retd. Home with me, at 11.30 p.m.

Feb 24. . . . All Fairs, Auction Sales, &c: prohibited to April from Feby 1st. Sadly inconvenient for Business . . . . Lambing – Ploughing Turnip land – Thrashing Oats.

Thousands of Rotten Sheep, in Lincolnshire, & Leicestershire – a *ruinous* affair!!!

March 1. Sharp Frost, could not plough.

Sir Geo. Colley and 300 others, killed & wounded by the Boers, in the Transvaal.[2]

March 10. Drove Jane to Witney – we Tea'd at Mr. Worley's. Wm. Shillingford owes £32,000 – and arranged to pay his creditors 2s.6d. in the £!!! . . .

March 12. Met Hounds at Dytchley Park – 16 Ladies out!!! Found 1st Fox in Oak Riding, after a *Fast* 25m. killed – Found 2nd in Kiddington after a *Fast* 20m. killed in Glympton and lost the Fox in Brook – found 3rd Fox in Glympton Gorse, to Ground in 24m. Home after a very nice afternoon's sport – met at 12.

Several mishaps at Glympton Brook, Miss Gaskell & several others down – and *in* it, I had water in both Boots – but luckily otherwise dry!!!

Prince Leopold, & Princess Wales, have each been at Cornbury Park.

March 13. Alexander 2nd Czar of Russia – assass[inate]d by a Bomb shell being thrown at him, which shattered his legs &c – he died 1½ hours afterwards 3.35 this afternoon – this is the 5th and last attempt on his life, since 1866.

March 21. Every prospect of *late* Vegetation, and *backward* Spring.

March 23. Attended Chipg. Norton Board Guardians & Ass[essmen]t Comtee. – Rent is Gross Estimated Rental, 10 per Cent off it, for Repairs, and the residue 'Rateable Value' – viz £1000 Rental £100 deduct for repairs, & £900 will be the Rateable Value of a Farm.

March 24. At Witney – Sold Wells 35 Tegs 59s. – and 34 for 63s. – lambs.

Vestry meeting at Fairspear House – Mr. W. & B. Hobbis, & Mr. C.A. Waller, attended.

Wheat 42s. to 47s. – Barley 24s. to 36s. – Beef 4s.8d. to 5s.4d. Mutton 6s.6d. to 7s.4d. Pork 11s.6d. score – Maize 30s. – Oats 1s. stone.

March 26. Mr. Bateman's Golden Wedding Day: of Witney, & late of Astally [Asthall Leigh]. Severe frosts *all* the week – only Beans sown!!!

March 30. Rode and joined Hounds at Tangley – killed 93rd Fox in Rissington in Revd. Rice's Drain.

Tea'd and paid small seed a/c to Mr. M. Matthews, Fifield – told me Godfrey, & Gillett, were both Insolvent!!!

April 2. Weather continues *most* severe, day & night.

Mr. Gillett, Tangley, & John Swain, Wrangle: liquidating.

Heythrop Hounds at Kiddington Gate for *last* meet of this season.

Mr. Mrs. Brassey & Family, start for their Yacht at Marseilles on Monday – go by London & Paris – visit various places in the Mediterranean, &c.

April 7. Mr. W.E. Gladstone introduced his Irish Land Bill, appears on the whole to be satisfactory to the Irish Members!!! which speaks volumes!!! if the Bill should be appreciated, it will prove the great crowning Act, of Mr. Gladstone's Political Life.[3]

April 8. . . . Drove over and tea'd with Mr. Savidge, Sarsden: told me sad tales about Earl Ducie's Tenants and Estates: letting his nice-sized Farms in 2 or 3 lots to poor men, unable to stock or work them – given Wm. Edgington a Cart Horse – to the chagrin of his neighbours, &c.

April 12. Earl Beaconsfield, dangerously ill, of Gout &c – begin to despair of his recovery.

Mr. Parnell is preparing the Irish, to consider Mr. Gladstone their greatest Friend!!!

Spent an hour with Mr. Haigh.

April 19. At 4½ a.m. Died the greatest Statesman of the present generation, Benj: Disraeli, Earl Beaconsfield, in the 77th year of his age.

April 21. Drove John Chas. to Mr. Gillett's Sale, Tangley: all sold off the place by order of the High Sheriff – Mr. G. upwards of 80 – been one of the best farmers in Oxon!!!

April 22. Rode to Charlbury, & attended Mr. Robt. Parkinson's first Auction Sale – Mr. P. will never excel as an Auctioneer!!! . . .

April 23. Mr. Maddox selling his Swede Tops – good idea, this *scarce* vegetable Spring: everything very backward, I have not a field of Spring Corn nicely up, even yet – not even the Beans!!! 80 acres Oats to Drill, & all the small seeds.

April 26. Earl Beaconsfield laid in a vault to-day by the side of his 33 years wife, in Hughendon Church – Prince Wales, Mr. Gladstone!!! was unfortunately prevented attending – Duke Connaught, & Prince Leopold, a great many of the Nobility – Foreign Ministers – with many Thousands of the late Earl's admirers – saw their unequalled Chief placed in his last resting place. . . .

April 30. Dined with Mr. Kimber, Hill, Revd. Bensley &c – at Mr. Brown's, Burford – Tea'd Mr. Garne's, with Jas. Shillingford, Mrs. Cheetle, and others – proved the *smallest* Burford Fair, on record.

Showers last 2 nights & this afternoon!!!

April has been a very Dry month!!!

May 3. Finished Drilling Oats in Leafield piece – considered quite *early* enough, for that description of land.

Attended Mr. Walton's Sale, Chalford, near Heythrop – bot. a Turnip Cutter 18s. and set Three [horse] Harness 7s.

Tea'd with Mr. Blake, Broadstone Hill: afterwards looked over the Shorthorns, Pigs, Ewes & lambs, & Ewe Tegs – 60 of which he had sold to an American for £6 each, after being clipped – Mr. B. chose the first 10 – and the American took the next 50. . . .

May 6. Finished Spring Corn Drilling – in Hall's Hill – Clipping Ewe Tegs – Rolling – Sowing small seeds – Horse-hoeing Beans, & stirring Fallows.

Very dry weather continues.

Cuckoo, Nightingale, Swallow, & Turtle Dove: all in full residence, amongst us. . . .

May 19. At Witney – and drove on to Mr. Eagle's, Northmoor, met a large party – killed 11 Dozen Rooks.

Mr. C. Simmonds, & C. Blake, from Oxford, & others gave us some good Songs – to bed 4½ a.m. to 7.

Heavy showers, all round the Country.

May 23. . . . Mr. Pettitt – a Crown tenant in Beds., cut his throat – thro' Agl. Depression!!!

May 24. Her Majesty 62 years old – Leafield Bells, rang at 4 a.m.

May 26. Very heavy rains, with *terrible* Thunder, at Enstone, Chipping Norton and Evesham – only nice mild gentle rain here, in the evening.

June 8. Jane and I, by 8.17 train – Paddington 10.30 – met by Mr. Mrs. Groves, Brockley Hill House, Stanmore: Middlesex – went to the Royal Academy – saw Sir John Astley and Earl Beaconsfield painted by Millais – &c – on to Horse Show – witnessed Mr. W. Kelk's 'Kathleen' jump Gate, Water, & Hurdles: we retd. by rail Home, with Mr. Mrs. Groves & Wm. Kelk to Brockley Hill.

June 10. Met Col. Sanderson at Horse Show – he showed us round Romano's & Short's, in the Strand: Covent Garden: &c – and we went to Fancy Fair at Albert Hall – saw Lady Churchill, Marchioness Blandford – Mr. Mrs. Samuda – Capt. Waller &c – inspected Albert Memorial – walked down Rotten Row and met with Mrs. Groves & Jane – tea'd at Bouthron's, and Mr. Kelk accd. me to Lyceum – saw the Cup – by Irving, & Miss Ellen Terry: beautifully performed.

June 11. Mrs. Groves started at 7.20 a.m. from Charg. X and would reach Paris in 8 hours 20 m.!!! we retd. to Town and Jane & I, drove to 48 Gresham Park, Brixton, after bidding Mr. Groves & Wm. Kelk, adieu: at

King's Cross, we retd. by St. Thomas' Hospital, Westminster Bridge – and attended an afternoon performance at the Opera Comique called Patience – which pleased us very much. Left Kings [Cross] with our luggage at 5.30 for Paddington – left at 6.30 for Charlbury at 8.39 – after spending 4 Busy Pleasurable days.

June 19. Mr. Ward, from Christ Church, Oxford: here at 9 a.m. Magnificent day – after fine showers yesterday evening. At Church in p.m.
Mr. Ward retd. to Oxford, by 7.15 p.m. train.

June 22 and 23. 18 hands flat hoeing in Kingswood, and Fernhills – single afterwards – 8s. per acre.

Jane and I, by 5.5 p.m. train for Worcester at 6.30 met with Mr. Humphries, who resides near Pershore.

Spent 2 hours at Theatre and saw 50 boys & girls sing, & perform – off about 10.30 and reached Edinburgh 7.45 – 10m. by rail to Portobello – Breakfast, with Mrs. Hockley and off to 'Auld Reekie'. Walked round, and had various fine views from Calton Hill – rather admired the 'Obelisk' – Inch Keith Rock in the Harbour . . . Admired Scott's monument in Princes Street. Walked up to the Castle, & views most extensive in all directions – down Cannon Street – saw Knox's grave – House – Houses Parliament – Judges sat in Session – St. Giles Cathedral & Holyrood Palace, Chapel remains, & a Beautiful Fountain in front of Palace – it is a cold, sad solemn, Prison-like Palace: saw where Rizzio was stabbed – dragged – & found dead &c.

Dined Portobello at 4. Walked on the shore and on the Pier in evening. . . .

June 25. Left Edinburgh 8.30 to Linlithgow – saw the Old Castle, on to Grahamstown, Larbert, Stirling, the Castle has a very high rocky site – from which you can see Ben Lomond, on to Bridge of Allan, beautiful scenery to Dunblane, where we stayed 30m. – Jane brought a Rose away to Fairspear!!! On to Callander – Coached hence to Trossachs Hotel – came by Lochs Venecar & Achray, and beautiful wild mountainous scenery – walked hence thro' Trossachs Valley by Loch Fye House to tail end of Loch Katrine – where we unfortunately lost our way . . . and had to return to Trossachs Hotel and wait: return of coaches – spoiled our day – missed seeing and sailing up Loch Katrine, down Loch Lomond – Mr. Ben Hockley was disgusted with himself – for the great mistake he made in walking us thro' the Trossachs, without thoroughly knowing the path.

Brought a piece of rock out of Trossachs valley to Oxon. Coached back to Callander, & retd. by rail, thro' same places as we passed in the morning – reached Portobello *11.15*, tired and weary, & disappointed!!!

June 27 and 28. To Edinburgh – drove round by Leith Docks, Whitehaven, saw the Fishermen's Wives & their Homes!!! on to Herriott's

Hospital – the Colleges – &c – walked into the Cathedral, a magnificent edifice – Sir Gilbert Scott, architect &c.

Inspected the Albert Statue & corner Figures. Saw the Gallery of Statuary – and Fine Arts Hall, which contained a fine collection of Paintings.

Left Waverley Station, Edinburgh, 4.25 train to Carlisle in 2 hours – stopped 40m. – stopped 10m. at Settle, admired Old Castleburgh Rock & enquired about the Robinsons – reached Worcester 4 a.m., left at 7 and reached Shipton 8.5.

Rode all round farm with John Charles, who is 21 to-day. . . .

June 30. . . . June has been a Dry month – a copious rain is much desired, and has been the past three months.

Summoned to Oxford Assizes on Monday next – as a Special Juror.

July 4. By 8.17 train to Oxford Assizes – waited till 4 p.m. before a Special Jury was sworn, Revd. Geo. Moore, Cowley v. Mr. Barling – adjourned 5.50 to 10 to-morrow.

July 5. By 8.17 to Assizes – commd. 10 – allowed 25m. & 30m., noon, and evening, for refreshment – and did not retire to give our verdict till 12 at night – retd. 12.15, each party to pay costs, and each culpable – witnessed most exciting scenes – and altogether much amused with the case.

Stayed night at 'Jones Railway Hotel'.

July 8. Guiteau, has confessed to shooting President Garfield: who at present is recovering, from his *serious* wounds.[4]

Herbert Marlow Ward, proposed to Alice.

July 16. Miss Susan Wilkinson, left us for London: after being our Governess for 2 years 8 months: Grace, Fred & Isabel, stood on the Garden wall, to see her go as far as they could for Ascott Stan. . . .

July 31. The heaviest amount of *water* fell last night, and this morning – since 17 and 18 Jany. last: wish it had fallen *middle* of May!!!

Aug 1. Recd. a letter from Whitehall Place to say £4000 of my Lease Rent, would be returned me, to assist and aid me, thro' the Depressed Seasons, for the *6* years ending Michaelmas 1884.[5]

Aug 5. School meeting at 7 p.m. – choosing a new Master: George Gordon – Messrs. Habgood, Bateman, Abrahams, Waller, & Self attended.

Aug 6. Walked round Mr. C.A. Waller's Farm in the evening, & a very bad one it is!!! 15s. per acre is the present Rent: and he is anxious to quit, at even that price.

Aug 12. Two days Wheat Reapered, and *all* hands ready for tieing – rainy morning!!! Rained & drizzled till midnight – Moon came forth – turned drying with a cold North wind.

Grouse are plentiful – but a many 'cheepers'.

Grace Catherine 13.

Aug 23. Rain, at intervals, all day: most wretched time for us. Very much corn cut, *all* over the Country – some of it been cut several weeks – and same sad story from all counties!!!

Aug 31. Carried 2nd crop Clover (*spoiled*) in Shakernook, and Kingstanding grass – for Beast & Horses.

Sept 16. Sunny fine day – the best for weeks!!!

Pulled nice crops of very fine Plums, & Apples off some of the Trees – and some liquid Manure and Nitrate Soda, forces them for bearing, both ought to be applied Spring & Autumn.

Sept 17. Began Harvest on 10th August: and only carried 8½ days!!! the best Harvest I ever grew, if I can only secure it in *good* condition – Wheats stained & much injured, & ricked in moderate condition.

Sept 19. . . . Pulled about 70 large Plums off the garden wall tree opposite Clements' House – the last ripe ones of the year.

John Wright Clinch, Witney: died suddenly at Fairford: Heart Disease: married Miss Townsend about 2 years ago – saw him at Brewery, when at Witney on 8th ult. Robt. Wm. & I, tasted the Ale, & saw the Yeast weighed.

Sept 23. On 19th ult, at 10.50 p.m. – President Garfield died from internal hæmorrhage, from a Pistol shot by Guiteau, inflicted 80 days previously!!! much and deeply regretted by probably all the civilised Countries in the World.

Our Court go into mourning for a week – for the first time in History – for the Ruler of the States!!! . . .

Sept 24. Carried corn 1½ days last week and 1 day and 2 afternoons this week!!! continuous wretched weather – for the carrying of the Harvest.

Rain at 3 p.m. . . .

Sept 26. Hired Richd. Gardner at Burford Fair – in Willett's place, as Carter.

Dined at J. Brown's, with Messrs. Reade, Porter, Hill, and Secker.

Sept 29. Robt. Wm. accd. me to Witney – we called upon Mr. Cole, and made enquiries about the Trade.

Wined with Mr. W. Worley.

Carrying Beans & Oats all day and *Finished* cutting: Oats & Barley in Hall's Hill!!! very glad – had a most tedious time!!!

The St. Michael's College opened, at Shipton.

Oct 1. Cool, fine, sunny day, after Fog till 9 a.m.

Carried all corn, except some coarse late Barley & Oats, which require *time* – before being carried.

Messrs. Reades, Price, Waller, tea'd and spent the evening.

The Marquis Waterford's Hounds forbid to Hunt, by the 'wretched Irish'.

Oct 3. Carried Harvest – except 10a. late Oats.

Fine dry cool day.

Mr. Clutton wrote giving the full charge of South Lawn, under my management.[6]

Oct 6. At Witney Michs. Fair and Mop.

Saw Mr. A. Bateman about Robt. Wm.

Oct 12. Hazleton & Hounds called here, on retg. Home, after meeting at Tangley.

Mr. Price accd. me to Witney: we dined with 47 others at Bailiff's Venison Feast – Revd. Foxley Norris in the chair &c. Home 10.30 p.m.

Oct 13. Carried 'Harvest Home' a small rick of late sown Oats, next yard, in Hall's Hill. . . .

Oct 14. A Terrific West Wind arose at 6 a.m. and continued for 12 hours – blew Thatch off Ricks, Sheds – a large Beech out of 'Fairspear Clump' and the largest Ash out of '4 Ash Bottom'.

Immence damage will be the result.

Oct 26. Very fine Autumn, for Wheat Drilling, Thrashing, Turnip Sheep, & general Farm work.

Nov 3. Mr. C.A. Waller has taken Hincksey Farm nr. Oxford, 411 acres, owned by Col. Harcourt.

Nov 4. Farmed round South Lawn with Mr. Waller, tea'd & spent evening. Mrs. Waller's 14th child!!!

Nov 14. Misty morning – Telegraphed Dr. Batt to Elizh. – Measles & Bronchitis!!!

Mr. Price and Jas. Reade junr. came to shoot Rabbits, we killed 69 and a Partridge.

Witney Feast *c.* 1895, taken by Henry Taunt. *(Oxfordshire County Libraries)*

Men carried 20 Carts load wood from 4 Ash Clump, so Adieu to 'Five Ash Bottom'.[7]

Nov 21. Isabel's 5th birthday – a great many presents – my Cocoa Nut – pleased her much!!!

Her Godfather Mr. Price, gave her a Florin.

Nov 24. Drove to Witney – and finally arranged that Robt. Wm. goes to Mr. Cole's on 2nd Jany. – for Learning Brewing, Malting, &c.

Gave Mr. Dale, Kings Head, Witney – 9 Jays & a Hawk for stuffing.

Dec 5. Leafield Temperance Society meeting – addressed by Robt. Sawyer, Maidenhead: Barrister at Law.

Rode South Lawn, and took down cropping and size of various pieces – from Mr. Waller's books.

Fine day – after mild drizly rain till 11 a.m.

Dec 10. Met Hounds at Mr. Adams, Hailey Manor, the Home for many years of the Famous Fox Hunter Assheton Smith – found in Coggs – lost after much running nr. Crawley – tried 'Stockley' blank – &c. . . .

Dec 12. Mr. Herbert Ward, retd. from College, via Fairspear. Farmed round South Lawn – attended Shipton Auction – sold 'Winceby' and 30 Fat Ewes – tea'd with Mr. Price.

Mr. Harvey Morris, Milton: found dead on his way home – from Wood Sale at Red Horse – Heart Disease – 48 years.

Dec 22. By rail to Oxford, to Marston Sale – bot. nothing!! Rode from Magdalen Bridge in Tram Car to Stan. for 2d. – first ride in them, & first sight of them, in Oxford.

Dec 25. Bright sun-rise, and dull, still day, with Frost. Our 7 sons and daughters dined together.

Dec 31. Dull misty morning – cleared up fine.

Great winds & storms, during the past year.

The Czar of Russia – blown to pieces – President Garfield, shot: and Earl Beaconsfield, died: during the past year!!!

Agriculture is sadly depressed – more abundant corn and Fruit & vegetable crops – but Harvest sadly injured, by frequent rains at gathering.

General Trade, is slowly increasing.

John Chas. shot a Peregrine Falcon, 42 inches across the wings – Ascott Com[mo]n Clump!!!! . . .

1882

Jan 1. At 8 a.m. our Groom, Jas. Sifford's son Frederic died ae. 8 – Measles, & Bronchitis.

100 to 200 children, at Leafield, ill from Measles: very mild open weather – *Turnips Clovers* and nearly everything, growing.

Jan 2. Drove Robt. Wm. to Mr. John Coles, Witney: tea'd, and signed Indenture with Messrs. Shillingford's binding for *One*, & Two years if satisfactory, to learn the Brewing & Malting Business.

Jan 3. Alice ill – Grace Measles – & Fred shaky & chill with every appearance of an attack.

Mr. Price, called.

Jan 5. Alice off to visit at Morville, nr. Bridgenorth, the Revd. Ward.

Midges dancing over Violets around the House!!!

Measles most prevalent all round – Grace and Fred – both in Bed – like Lobsters!!! . . .

Jan 12. Drove Elizh. Anne to Witney.

Grace & Fredc. – recovering from what Isabel calls, the 'Weasells' – measles. . . .

Jan 19. Rode to Woodstock, and heard Earl Lytton & Sir H.D. Wolff address the Conservatives – Lord Randolph Churchill in the Chair.

Feb 1. . . . Two men looking over South Lawn from Birmingham. Fine, clear morning, with slight black Frost: after one of the *mildest, driest*, Januaries, on record!!!

Feb 3. Met Hounds, Farmington: large 'Field' – ran brace Foxes – eventually lost both – Travers the Cotswold Huntsman out – Mr. Purse, Master of the Galway Hounds visiting Heythrop: all the Hounds, but 2 Packs, are prohibited Hunting by the Irish *Fenian Land Leaguers*!!!

Feb 4. John Chas. met Hounds at Wilcote – did find at 3 p.m. in Pinsley. Luncheoned at Mr. Sutton's.

I farmed round Home, and South Lawn.

Finished manuring the young Clovers in Ascott Common & Hall's Hill.

Only 15a. Turnip land to plough to catch Sheep.

Magnificent weather for Agriculture!!!

A. Brassey Esqre. given £500 to the Irish Landlord Property Protection Fund.

J. Chas. at Leafield Temperance meeting.

Feb 25. Finished 40a. Beans drilling in Leafield piece – Thrashed Wheat, Barley, & Oats: 4 days – Ewes began lambing yesterday.

Met Hounds Eynsham Hall, had Biscuits and Champagne & Cherry Brandy – found in Coggs, & lost towards Witney: tried Pinsley and all the hanging Covers to Wilcote – which was also blank – Sherry at Mr. Sartoris', & Home. . . .

March 2. Wild, cold, showery day.

'Maclean' shot at the Queen – fortunately neither hurt or even frightened her – the Bullet struck a Railway Truck – Her Majesty was being driven from Windsor Stan. to the Castle: the 7th & I hope the last time, anyone will make her, a target.

March 6. Attended Mr. Morris' Sale, N Leigh – bot. Horse Rake & History of England.

Tea'd with Mr. Bryan Stratton Dean &c at Mr. Sutton's, Wilcote.

March 22. Ground covered with snow till noon –

At Mr. Garne's Sale, Broadmoor, bot. 21 ewes, 26 lambs £106.19.0d. for South Lawn.

Mr. Garne's Family comd. Farming Broadmoor 670 acres under Lord Sherborne in 1824 – the place has lost them £9000 the last 6 years.

Attended Church at 7 p.m. – heard Revd. Horlock who is going out to South Africa.

March 31. Farmed round South Lawn – attended Mr. Jesse Cook's Sale, at Barley Park – and had an hour at Witney.

March been a very fine Month, for Agriculture.

April 24. Very busy at South Lawn – repairing Gates, iron Hurdles, Buildings, Cottages – cleaning the land before sowing with Spring Corn: new stone wall half round the Cottages nearest Shipton – 18s. per chain. . . .

April 25. The Bishop of Oxford confirmed about 150 at Leafield 11 a.m. – Rain from 12 to 6 p.m. . . .

April 27. . . . Mr. Gladstone's Budget, has raised my waggonette 21s. and Pony carriage 6s. – so much for the relief of 'Agriculture'.

April 29. . . . Hauling Manure – too wet for even stirring Fallows. Ewes and Lambs finished Turnips this day – never had them last so long for 30 years!!!! . . .

April 30. The first half of month – dry: the remainder sadly wet, Plough stopt, even on Fallows!!!

Mr. H. Ward, spent the day from Oxford.

May 5. Mr. Gladstone sadly in trouble with Ireland – Earl Cowper the Lord Lieut. – Mr. Forster, Secretary, &c – all resigned.

May 7. Lord Frederic Cavendish and Mr. Burke, murdered, stabbed to Death, in Phoenix Park, Dublin 7 p.m.: the 4 Villains drove away unpursued by anyone!!![1]

May 11. An hour at Witney – forward & reached Mr. Eagle's Stanton Harcourt [at] 5: went up Pope's Tower round Major Wynter's Garden, admired the Wild Duck and Water-hen nests & brot. away an Apple out of Pope's Study – killed 6 Dozen Rooks – Revd. Walsh, Dr. Batt, C. Simmonds, Martin – Hanley – Walker, Bateman, Blake, 2 Mr. Bulfords: &c. Home at 1 a.m.

May 15. Mr. Jeffery's Cultivator comd. on South Lawn Fallows cultivate twice over 14s. and Drag 4s. per acre.

Only suckling ewes to clip – or shear!!!

Not so many Swallows, Nightingales, Turtle Doves, or Cuckoos, as usual.

May 17. The coming Harvest will be most various – spring corn to drill even yet – and the Wheats *earlier*, than known for some years – Meadows, & Clovers, same.

Cold, dry, sunny weather.

May 24. The Queen, my Noble Landlady: 63.

Drove Jane to Agl. Show, Witney: called at Mr. Coles. Tea'd at Mr. Westall's – stayed Concert, Corn Exchange. . . .

May 25. Rained 3 hours at Witney yesterday morning, and 6 to 9 hours to-day: consequently spoiled the Show – as hundreds were entirely deterred!!! been quite a Curse on Agl. Shows, past few years, & occurred in the most wretched of Weather!!!

Ordered 20 tons 'Best Super' from Mr. John Robinson, Bristol – £3.17.6d. per ton – Disct. off to July 31st.

June 6. At 5 p.m. – Dr. Haigh, Scovell, Brooks & Franklin called & I accd. them to an inaugural Banquet held in Witney Corn Exchange, 300 dined, W.A. Hall presided, supported by Earl Jersey, E.W. Harcourt, Staveley

Hill, Risley, Norris, &c.

Tea'd at A. Stratton's 1 a.m. as we returned – Home 2½ a.m. – all passed off remarkably satisfactory.

June 7. Mr. Mather quits Chadlington at Michs.

Earl Ducie's Steward, Rich, quits at Michs. to the great delight of all the Tenantry!!!

June 8. Jane and I, Alice & Mr. Price, went to Ascot Races, 'Foxall' won the Cup – the very worst course I ever saw, for seeing the Racing. We called at 'Comins', saw Royal Buck Hounds & had Lunch at Head Whip Comins's – walked from his House by the Huntsman, Goodall's – and took a stand on Course beside the Furzes, and had a very good view of Royal Procession.

The Prince Wales, & Princess, & Princess Helen, looked well, also saw the Duke Cambridge.

June 10. By 1.15 train and met Mr. C.A. Waller, and finally settled *all* up, respecting South Lawn, at the Roebuck, Oxford.

Mr. Herbert Ward accd. me Home.

Very showery unsettled weather – cannot drill Turnips, and dare not cut Clovers.

June 20. Victoria ruled over Old England 45 years today. Edward 3rd reigned 50 – Henry 3rd 56 years and George 3rd 60 years, & for number of years her present Majesty comes next – and may she beat them all – she is hale, & strong enough, to do so, at present.

Mr. Price brought us a Salmon, sent over from Ireland, by Sir Morgan Crofton Bart.

June 23. Fine sunny morn – after a very copious Rain – 120a. Clover just mown – 60 acres Turnips in – and will soon be 'out'.

Clover 37a. cut at South Lawn, & 60a. Turnips in, there.

June 25. At Church in a.m. Dull cheerless day.

Wonderfully fine growing weather – the Country full of Vegetation, every prospect of abundance – if we can only be blessed with a fair amount of *Sunshine*!!!

June 28. Drove Chipg. Norton, & attended Board Guardians &c. Lunch with Mr. Savidge – told me about Mrs. Sartoris & George Dawkins – the *terrible* state of the Ducie estates – tenants – Steward, Rich, &c.

June 30. June has been a dull, rainy, cold, blighty month: some of the Oaks, Black Thorn, Hawthorn, &c: are leafless!!!

July 1. Had 3 fine days for Clover carrying – which has relieved all anxiety – 126a. here, & 37a. cut, at South Lawn. . . .

July 8. Met the Hon. Chas. Gore, & Mr. Clutton, at South Lawn – drove and walked all round the Farm: rain set in heavily at noon, and prevented their driving round my farm. . . .

July 11. Comd. bombarding the Forts of Alexandria – the first trial of our Big Guns.[2]

By 8.15 to Reading – Royal Agl. Show – lunch at Suttons, heard Dukes Richmond, & Buccleugh, Lords Vernon & Galway, Dent &c. – rain set in & I left washed out at 3.20, Ascott at 5. Tea with Revd. Yorke, & Home in torrents of rain.

July 15. Saint Swithin's – rainy night & Morning!!! 40 days rain – will spoil *everything*, & *ruin* Farmers. I have 35a. Clover spoiling daily – cannot clean or drill land, with Turnips – ploughs mauling Fallows.

Drove Miss Nellie Bullock & Lizzie to Mr. Worley's Brize Norton – bot. 4 Rams 7 Gs. av.

Mr. Mrs. Hughes, Mr. Habgood, &c: tea'd: Home 11.20.

July 17. Mr. Dan Westall & his sister, Tennis & Tea.

Samuda's Garden Party at Shipton Court.

July 20. Yesterday & to-day finished carrying Clover out of Hall's Hill, Witley Hill, & Ascott Common, after being delayed by *Rains & Showers*, since Wednes. morn. the *5th ult.*

July 21. 13 men mowing Charlick off, last sown Oats, on South Lawn.

Drove Jane, Eliz., Grace, & Fred., & Isabel, to South Lawn: for Currants, Rasps, Strawberries, & Cucumbers.

Drove John Chas. to Wilcote: Tea'd with Mr. Mrs. Sutton, looked over Mr. C. Sartoris, Garden &c: tasted the Strawberries, & Figs.

We walked on to Mr. Geo. Gask's Bridewell Farm, overlooked new Buildings &c – had a deal of talk about Linc. Friends, & their Farming.

July 22. Mr. Mrs. John Worley junr. Brize Norton drove over & spent the evening – shot himself couple Rabbits.

Mr. John Addy, looked over South Lawn, and appears a suitable man for the place: he resides at Stamford: Linc.

July 31. . . . July been a wet ugly month – for cleaning land, carrying Clover – or drilling Turnips.

Aug 5. . . . A very large Gull caught amongst Ducks at Kingstanding, sent it Mr. Arnott, St. Harcourt, for stuffing.

A large Fire Balloon alighted in Hall's hill at 8½ p.m. to the surprise of us natives.

Had a fairly dry week at last: after 3 months, wretchedly miserable weather.

Aug 7. Drove to Witney – & trained to Fairford and Bus to Cirencester great Sheep & Ram Fair.

Mr. Hobbs, Maiseyhampton: averaged £16.0.7d. for 50. Tea'd at Mr. Hobbs as we retd. to Fairford Stan.

Called at Mrs. Worley's, Witney.

Aug 9. By 8.15 to Oxford Ram Fair – bot. 2 Mr. Gillett's G11 & 10. Called upon Mr. Mrs. Pincham – Mr. Price & I, returned by Witney – tea'd Mr. Abraham's, Ringwood: after which he drove us both to Fairspear – where Eliz. Anne was entertaining her School Form of Girls.

Aug 17. At Witney – 2 Reapers & all hands on – for the first whole day, in *thorough good* condition.

Mr. R. Eagle, 15 2 yr. 4 m. old Steers from South Lawn, at £20 each.

Aug 19. Sir Garnet Wolseley, is daily expected to attack Arabi Pasha, & his numerous Rebel Army, near Alexandria – expect he will attack on 3 points at once, so as to prevent Arabi's escape, unless he faces the Libyan Desert??!!

Only been able to cut Corn 1½ day with Reapers, this wretched Harvest week: cutting is general all over England, & crops various, more so than usual, some districts have suffered from frequent Rains for 3 months (we have around here) and scarcely any Rain has fallen for that period around Cambridge, the East Coast, &c.

Aug 21. Yesterday evening attended Swinbrook Church, some of the Finest Monuments I ever saw, of the Fettiplace Family.

Mr. G.W. Shillingford came & weighed Wool up, 312 Tod, send it into Granary at Eynsham – because damp in wall was injuring it, in my place.

Mr. Taylor, Turk Dean, kept a/c for Mr. Shillingford.

Aug 24. At Witney – bot. Certificate – witnessed Reaper finish Cockshot-hill Wheat at 7 p.m. – shot 6 Hares 4 Phts. & Landrail – 75 Pheasants, 20 Hares &c amongst the Wheat, shot the 4 fine young Pheasants merely for a taste, because they all go into Cornbury Forest and I never have a chance of them again, altho' bred & fed, on my Farm – in shooting the 4 Pht. a Landrail was accidentally winged: many of the Hares return to forest when corn is cut, & *feed* on Farm every night.

Aug 26. Capt. Wilkins, & Chipg. Norton Volunteers, marched to Shipton, paraded in Play Paddock and were entertained by Mr. Reade – I carved, &c.

Aug 27. Sir Garnet Wolseley and 22,000 others are gradually and surely, taking in hand, the whole of Egypt; they now have Alexandria, Port Said, Ismalia, Suez, and all the Canal under control and Cairo next, expect a Battle between Dagazig, Tantah, or Kafr Dwor: our Big Guns and Marines have worked wonders, both on Land, Sea & Rail; we never utilised *fighting Trains* before – they used a 40 lb. Cannon, Nordenfeldt Guns & Gatling Guns – Sir Wm. Armstrong's Cannon, have proved the hardest hitters. . . .

Aug 30. Dry morn: 8 Wagns. & 3 Carts, will be on, every fine day, 300a. ready to Rick!!!

Sept 5. Drizly morning & rainy afternoon – a splendid rainbow 6 a.m. Harvest in a terrible pickle.

Sept 6. Mr. John Addy, & wife & mother: looking over their future new Home of South Lawn.

Mr. James came from Whitehall: also met Mr. Groves and arranged for Painting House &c.

Sept 9. Carrying 8 Wagns. 3 Carts – 1 Reaper cutting Oats. Several hands tieing Oats, & cutting Fernhills Barley, & Leafield Piece Beans.

Sept 18. Very fine day, carrying 9 Wagns. 1 Reaper at Beans, & Thrashing Wheat – 4 Thatchers on.

Sept 19. Attended Mr. Sargent's Stock & Implmt. Sale, at Sarsden Mount. Rain at noon.

Sir Garnet Wolseley attacked Arabi, & Toulba Pashas, early on Wednesday morning last – unexpectedly!!! drove them out of Tel-el-Kebir Entrenchments: put them to Flight, ordered another party to march on Cairo, which surrendered – Arabi & Toulba, taken prisoners, & the Egyptian occupation is virtually over, to the great Glory of our Navy & Army, in all its branches: who did everything they attempted in the most brave manner.

The Duke of Connaught, & Sir Garnet W. are Palaced in Cairo.

Sept 20. Rainy day – hired all Labourers &c – Clement leaves for Farming at Leafield: Wm. Painting takes his place.

Labourers House & Garden 1s. rent per week – Carters 13s. per week – House & Garden – 20s. for Ale in Haytime & Harvest, & 30s. Michs.[3]

Oct 1. Drizly mild morn – hundreds of acres of Corn to cut and carry around here: only carried 3 days the last fortnight.

Oct 9. Mr. Clutton wrote me for Harvest Report!!!

Drove to Eynsham Hall – met about 30 for Wild duck Shooting – we shot 126 Mallards, 10 Black-green Indian Ducks, 4 Moorhens, 1 Coot, 2

Rabbits. Lunch and Champagne in profusion in the Gymnasium. Some very good songs – Hunt, 'this day a Stag must die' I thought the best song of the evening.

Oct 11. Painter, came as Foreman.

Met Messrs. Addy, Davids, & Richards, at South Lawn for valuing: showery wretched day. Called at Shipton Lodge as I returned – reached Home terribly drenched!!!

Mr. Lockwood preached 'Harvest Thanksgiving'!!! Sermon at Leafield: 23a. Oats, and 40a. Beans spoiling in the Fields.

Oct 13. Fog Drizzle & Rain 24 hours – Barley, Oats, & Beans – growing & rotting on the land – cannot Thrash, carry Oats (23a.) or Beans (40a.) nor Drill Wheat, or even haul manure, on Seeds!!!

Mr. Jas. Mace, Solr., Chipg. Norton & Mr. Prior called on their way from Ringwood – and took my Depositions for appearing next Tuesday at Sessions, Oxford, against the Assessment Comtee. of Witney Union: on behalf of Mr. Abraham's Assessment.

Sent in, the last Crown Labour a/c's on South Lawn – a few minor bills, and all will be finished?? and glad I am – the Weather is most disheartening!!!

Oct 16. Sent Mr. Barford, Banbury: Mr. W.E. Mawer's Geraniums – 20 per cent amongst the White Clover Seed!!! the seed resembles the Clover seed – hence the adulteration on us ignorant Farmers!!!??

Oct 17. . . . By 8.13 train to Oxford Quarter Sessions, as witness as to the gross rental of Mr. Robt. Abraham's Ringwood Farm – Messrs. Jonas Paxton, Scroggs, Timms, Savidge, Bateman, &c: described, and gave their Value of the land – Mr. Thornhill: Chairman – case dismissed – each party pay own costs – which was equal to a Win for Mr. Abraham: as against the Omnipotence of an Assessment Comtee.

Mr. Abraham entertained us all at the Roebuck, and we retd. by 4.25 train.

Oct 18. Mr. Matth: Savidge, Sarsden Lodge: has given Earl Ducie 'Notice to Quit' – his Lordship would not reduce Mr. Savidge's Rent. . . .

Oct 21. Rain, Rain – Thousands acres corn growing & rotting in the Fields, between here & Cheltenham: I have 24a. Oats & 40a. Beans to carry – only carried 3 days, & 2 parts days – *five weeks*, ending to-night.

Wrote Mr. Clutton, the state of matters!!! . . .

Oct 27. Not carried Oats & Beans for 3 weeks – nor been able to Thrash, or Drill Wheat!!!

Mr. John Addy, reached South Lawn, for permanent residence.

Wind and Rain!!!

Oct 30. Opening day at Heythrop.
Sorted Tegs – $\frac{3}{4}$ lb. old split Beans to the best & worst – $\frac{1}{4}$ lb. to stock Tegs – & 1 lb. each to Feeding Ewes.
Mr. John Addy called & bot. Charlie's 12 Pigs 20s. each 8 weeks old.
Rain!!! & Showers.

Nov 6. Only drilled Wheat, 3 days – Oats & Beans growing & rotting, in the Fields – carried no corn since Friday 6th October – altogether the most ruinous prospect for Farming I ever experienced.

Nov 13. Carrying Beans, after a delay of 37 days, thro' frequent Rains, Snow, & Fogs: the 6th. Oct ult., the last corn carrying day for miles around here. . . .

Nov 15. Carried 'Harvest Home' began 10th Augt.!!!

Nov 22. Dull day, after a rainy night.
Grand Review on Saturday last in London – of all those who had passed thro' the Egyptian Campaign: the Queen made a special *rising* Bow, when Sir Garnet Wolseley, passed her!!!

Nov 30. At Witney – very large market – Mr. J.D. Bliss made 15s.6d. each of Tegs.
Saw Mr. Flood's 15th Pike: would not cash Gask's cheque.
Mr. J. Addy, & Brother, there.
This has been a *Wet* Month: 11th to 15th the *only chance* to finish carrying Harvest!!!

Dec 1. Finished Wheat drilling – what a struggle it has been, surely.
Met Hounds at Bourton Wood, 17 miles hence – the only 'meet' and cover I had not seen in our Hunt – it belongs Earl Redesdale, who hunted with us *all* day, altho' upwards 70 years old – killed 1st Fox in wood – ran 2nd to Seizincote, Longboro', Donnington, Broadwell, & lost beyond Stowe, after running round Mr. Sartoris' at Swell: I reached Home at 4.15 after seeing a deal of the Country.

Dec 8. 3 inches Snow – Rode to Charlbury: & met Surveyor Taxes, Comber, & Revd. West, & Carpenter: appealed against £40 House Duty, & Schedule B £1320 at 3$\frac{1}{4}$d. in £.[4]
The Alhambra Theatre burnt down: Leicester Square.
A train lost in the snow, nr. Bala, North Wales.

Dec 14. Snow, Fog, & cold thaw – can do *nothing*, but haul manure.
John Charles at Witney – instead of me.
'50 years, as yesterday', our Premier, Wm. Ewart Gladstone, was returned as M.P. for Newark!!!

Manifold indeed, have been his Political Works, deeds, & changes, in the Half Century!!!

Dec 19. Wrote Mr. Clutton, after Mr. Price, Jane, John Chas., Fred, Grace, Isabel & Miss Dolman had gone to a concert at Leafield: John Chas. played the Violin to Miss Dolman on the Piano. . . .

Dec 20. Mr. Bliss, Waterman's Lodge, died last Friday night – ae 44: buried at Leafield this afternoon – 10 children: injured, many years ago, by a Horse running away with him, at one of the Forest Fairs – broke his Thigh, &c – always found him an agreeable neighbour. . . .

Ploughing for Barley in Big Brazil, & for Beans in Boynal Gardens. Digging foundations for Stone walls in, and across Kingswood, and Mr. Groves' men are making Wall across Cockshot-hill, and between Witley Hill and Fernhills.

Dec 21. All sheep on Turnips.

At Witney – arranged with Mr. Coles, that Robt. Wm. remains another year. . . .

Dec 31. Foggy night – with rain at 7 a.m. till night, then Fog!!! Been wet, with very little Sunshine, since 20th April, last!!!

The past year, has been famous for, *Terror, Murders*, and Lawlessness in Ireland – The Closure in Parliament[5] – The Egyptian Army Revolt – Cetewayo's return to Zululand – &c.

Been one of the *most hindering* years for Farming, I ever experienced – from *20th April last.* . . .

1883

Jan 3. Drove Jane & Fred. Geo. to Witney.

Advanced a further £50 as Premium to Messrs. Shillingford & Co. for Robt. Wm.'s 2nd year.

Tea'd at Mr. A. Bateman's.

Jan 11. Drove Grace, & Fred, to Witney: *Foggy*, ugly, *cold, Depressing* weather!!!

Tegs & Ewes great prices – quite 1s. lb. for Tegs in Wool – wish Wheat Oats & Barley were 10s. per qr. higher, & sheep 10s. each less, in value: Pork only 6½d. per lb.

Jan 13. Dull Foggy day as usual – can do nothing Dry.

Mr. Early's 'New Mills' burnt down last night. Estimated loss, £15,000: they stood a mile from Witney.

Jan 23. Dined 5½ p.m. at Mr. Richd. Eagle's, Stanton Harcourt: with Revd. Walsh – Majors Wynter & Booth, Bulford, Willesden, Worley, Adams, Jeffreys, Bryan, &c: we enjoyed a *rare* dish!!! a Haunch of roast 3 shear Hampshire Down Wether – ate like Venison: Home 1.10 a.m.

Drove up to New Mills Debris.

Jan 30. Mr. John Rowland, John Chas., Alice, Eliz. & Miss Hanger, played Piano, Violin, & sang, at Leafield Concert.

Feb 12. Shipton Auction – very windy rainy day: sold 15 Tegs. All sheep offered, to be 'marked', and killed in 6 days: thro' Foot & Mouth Disease, having broken out, in various parts of the Country.

Tea'd with Mr. Keck, at Mr. Price's.

Feb 13. Met Hounds at Wolford Inn, then found in wood, and ran to ground ½ mile from Mr. James Harbidge's L. Compton. Hounds went for Whichford Wood – I left and called Mr. Harbidge's, admired his Deer – and curiosities in the House – Sherry at Revd. Lockwood's and Home at 4.15 p.m.

Alice & Eliz: to London, by 8.13 train a.m.

Feb 26. . . . Mr. Wm. Bliss, Chipg. Norton, died this day ae 73. A Famous Manufacturer of Tweeds, Shawls, &c: worked 800 hands: left his 2 daughters £20,000 each, & his only son the residue.

March 8. John Chas. accd. Mr. Price to Moreton Steeple Chases – lunched with Mr. A. Brassey – Mr. H.A. Brassey – and Mr. Byass: which was about the best part of it – snow squalls, wind, & Frosty cold air, were in great force.

March 10. Turnips done, except 20a. White Swedes in Shipton piece on which there are 82 Ewes & lambs – Stock Tegs in Shakernook – 42 in Langley for Sale – 20 Ewes died from decayed lambs – thro' laying in mud &c – on Turnip land – hear of great losses from same cause.

Saw Shipton Auction newly bricked sheep pens.

Tea'd Mr. Price.

March 12. Shipton Auction of sheep sold on Brick bottomed pens – Sheep all branded with a large M – all to be killed in 6 days!!! the worst Trade of the Season!!!

Mr. Gask & Son there, sold some Scotch Highland Cheviot cross bred Sheep: I showed him round Shipton Avenue with Mr. Reade.

March 18. Mr. Marmaduke Matthews, Fifield: died on 13th ult. age 69 – a very respectable Farmer – he once refused the Duke of Beaufort, riding over his Wheat field – His Grace asked if he knew who he was – yes, said Matthews, you are the Duke of Beaufort – I am 'Duke Matthews!!!

March 28. Met Hounds Barrington Park – found over Farmington Grove, ran the Clapton Vale, Bourton, thro' Farmington Grove, over to Sherborne Park, turned down for Cow-pasture – over Broadmoor grass, & lost nr. Clapton Gorse – very nice run, several falls – we were photo'd at 'Meet'.

March 29. Attended a very clever lecture on Agriculture, by John Buckmaster, 200 attended.

Tea'd with Mr. Townsend, Price, A.J. Matthews & Wm. Shillingford, at Mr. James Shillingford's.

April 5. Only 18a. to 20a. Barley to Drill – Weather magnificent!!!

Mr. Herbert Ward, placing a Wooden Seat around the Oak on Lawn – & contracting a severe cold [at] same time.

April 13. Finished all Spring Corn in – Wheats rolled – and part of the Clover for mowing; and 40a. Fallows stirred, never so forward with work on the Land since my arrival here 7 years ago!!!

April 16. Mr. Eagle came & bot. Mr. Price's 8 Bullocks, & 2 Heifers, £21 average.

Jane to Witney shopping – Isabel bot. a handkerchief for 3¼d. gave the Draper 1s. began to think him slow in giving her the change!!

April 17. All work finished at present – except stirring Fallows!!!

Mr. Groves placing down, 2 Hunting Gates, in Shakernook – 95 chains new Walls entirely completed!!! 3 ft. 6 high & Toppers, 24 inches foundation runs to 15 inches on which to place the Toppers – Witley Hill wall, 6 inches higher, & 2 inches thicker for 35s. per chain – the others 33s.6d. Mr. Groves digs foundations – makes up Quarries – after digging the Stone – *hauls*, & builds!!! cheap.

Attended our Conservative Assocn. Dinner, at Shipton, Revd. York, Messrs. Kent, Addy, Stapleton, Reade, Spackman &c made some very nice speeches, altogether a most pleasant enjoyable meeting.

April 23. Rode with Mr. Price and called to ask after Mr. Eve at Sandford – found him suffering from Diseased Heart, Pleurisy Lungs, & enlarged Liver!!! almost sure to die, a hard, dark, hale, sober man, looked like living half a century, half a year ago.

Tea'd on our return at Mr. Midgley's, Chadlington.

Frequent Storms of hail from the North, & very cold. A terrible deal of Foot and Mouth Disease amongst the Sheep and Beast, all over Lincolnshire – will sadly affect the Trade this *whole week*, at Lincoln Fair.

April 25. Very ill – cold on a disordered stomach – sick & purged 24 hours!!! hope this will prove a cure till Autumn!!! as it generally has done previously.

April 30. Been a fine dry Month for Agriculture – work not been so forward for the past *10 years*!!!

May 2. By 8.15 a.m. train to Oxford: with Alice & Eliz.

Prince Wales laid foundation stone of Indian Institute – Lunched at Balliol – &c: left 4.20 for London – after a pleasant peep, at one of the happiest places, of his youth.

Saw Sir Thos. & Lady Brassey, & his eldest Son, Alnutt.

May 6. Mr. Abbey, Oxford: gave a Temperance Lecture last evening at Leafield – Bread versus Beer!!!

Lord Fredc. Cavendish & Mr. Burke, were murdered a year to-day in Phoenix Park, Dublin – their murderers Joe Brady, & Kelly, are under Sentence of Death on present date.

May 7. Showers at 11 a.m.

Rode over, & farmed round, with Mr. Matthew Savidge, Sarsden Lodge, had bot. a Bull of Sir Henry Allsopp, nr. Worcester – 35 G.

Mr. Craddock, had regaled the School Inspector with Port, to the benefit of Parochial Affairs!!!

Only 9 inches Water in Pump House?!!! Very little rain fallen the past 3 months!!!

May 9. Bot. Æsop's Fables for Isabel.

Lunched with Mr. Mrs. Clinch, Witney: also looked over the Cemetary as I rode by – & saw the Graves of late John Wright Clinch, & Dr. Augustus Batt – &c. 'Sic Transit Gloria Mundum'.

Mr. R. Price drove Cymro and attended late Mr. Chas. Eve's funeral at Sandford.

May 12. Rainy night, & morning: a Soaker, at last: only 9 inches water in Pump House: had but little rain, since first 10 days Feby.

John Chas. to Oxford – and on to Mr. Worley's, Brize Norton: Rook shooting: 8 dozen only.

Stowe on the Wold Fair.

May 14. Rode down to Shipton and inspected 1037 sheep – 8d. & 9d. lb. – had Lunch with Mr. Price.

Drove to Mr. Eagle's, Stanton Harcourt: and met Mr. Bulford, Hordley, to shoot Rooks at Major Wynter's, his Brother & he helped us – Mr. Eagle & Son Blakeman joined us to Dine at 8 p.m. – the Major showed me Shields, a Knob-Kerry, Damascus Sword, Buffalo heads, Dagger Knives & several other Trophies from Kandahar &c – altogether had a very enjoyable evening – reached Home 2.50 a.m.

May 28. Attended meeting at Crown Hotel, Shipton – and arranged about Subscriptions for 'Beaconsfield Hall'.

Drove Mr. Price and tea'd & spent evening at Mr. Robt. Abraham's Ringwood House – some of the Chestnuts in Garden had already grown, 16 to 22 inches long – lots of small Walnuts visible.

May 29. Rode to Charlbury and joined Earth-Stoppers party for 2 to 3 hours – 39 had Dined at the Rose & Crown, besides Hazelton the Huntsman and Allen the 2nd Whip – some very good songs.

Called at Mr. Bolton's, Finstock on my Return.

Dr. Haigh called 9 p.m.

May 30. Mr. Price bot. some Salmon – splendid weather.

Mr. F. Westall, Mr. Haigh, & Mr. Price in the evening. Mr. Westall told me, he held about 57 Inquests per year, in the Witney district – he much admired the Langley Oak – and Big Nut Tree, in Shrubbery.

May 31. May has been most favourable to Vegetation, & been free from Frosts, & especially 'Blights'!!!

Moscow very gay, with Coronation of Czar – there are 5000 Bells in the place.

June 3. Fine sunny morning, with brisk wind from East *all* day, but mild!!!

Rode down to hear Mr. Yorke preach the last time at Ascott, before he enters upon Fifield – stayed Sacrament – at Leafield Church in afternoon, and Jane, Eliz., Grace, & Miss Hanger, were driven to Ascott Church.

June 5. Jane & I, by 8.15 train reached Ascot, Berks. Reached Royal Stag Hound Kennels just in time to see Goodall, Comins, &c: start away to head the Royal Procession – breezy fine dusty day – Geheimniss won the Trial Stakes – Offspring, the Maiden Plate – Border Minstrel, the Gold Vase – Galliard, the Prince Wales Stakes – Superba, the Biennial Stakes, and Ishmael the Ascot Stakes.

June 6. Mr. Abraham drove Mr. Toller over, farmed me round and spent the evening – Mr. Price joined them: quite approved of the new Walls – improves the appearance, as well as being useful & serviceable, for the occupation of land.

June 28. Coronation Day – John Chas. Birthday.

A. Brassey Esqre. paid Earl Redesdale, £1000, for the Heythrop Pack of Foxhounds: they were valued by Earl Macclesfield & Visct. Galway: 70 Couples.

June 30. This has been a fine, Showery, Thundery, growing month and this Country is looking remarkably well.

Elizh. Anne retd. from spending a fortnight with the Bullocks at Bristol.

A French party have started for the Red Sea – in search of Pharaoh's Treasures, war implements, perhaps Chariots, &c which they expect to find embedded at bottom Sea – I hope they may succeed?!!

July 4. Drove Jane Witney – About 2300 fleeces wool – offered by Habgood & Sons – Longwools 20s. – 21s.6d. – Down 25s. to 26s.6d. per tod.

July 5. John Winston Spencer-Churchill, 7th Duke Marlboro', died last night of Angina Pectoris, after being Duke *exactly* 26 years: there are 2 Sons, & 6 Daughters: only Lady Sarah unmarried.

July 9. Pulled a peck Cherries – also a nice lot Strawberries.

Shipton Auction – John Chas. tea'd Mr. Reade's, with Mr. Abraham.

Painting exterior of House, Cottages, & Buildings: has much improved their appearance, *10 years* since done!!!

July 10. Drove to Blenheim Palace, Luncheoned in Palace with the Corporation of Woodstock, & the Tenantry, received Gloves & stood on the steps at Front Entrance Door, & witnessed the Coffin of 7th Duke Marlborough brot. shoulder height, and interred in Chapel, a very solemn, imposing sight – there were upwards of 25 Dukes, Earls, & Marquises, followed in the Procession, the late Duke's 6 Daughters, stood behind the Door near Organ – they will probably never be there together again in Blenheim Palace Chapel!!!. . . .

July 14. Began only, and carried Clover till noon, when heavy Thunder rain set in, not had one whole fine day all the week – most annoying with cleaning Land? or Haymaking?!!!

July 19. Finished carrying 1st cut, of about 75a. Clover at 3s.

July 20. . . . Showers at 10 a.m. – and heavy rain at noon – most grievous Hay-making weather – and all the heaviest crops Corn, & Potatoes, will be injured: unless we have a speedy alteration of Dry – or Sunshine, in the weather!!!

July 25. Threshing Oats carried last Novr. – very much injured – Straw only fit for Litter!!!

July 27. Drove Mr. Price to Mr. Garne's Ram Sale at Aldsworth: 51 Cotswold Rams averaged £20.1.6d. – four of which were 'Let' for the Season, at 85, 21, 15, and 17 Guineas – very large respectable Company.

July 31. This has been a sunless, rainy, cold, showery month: cannot clean Fallows, or the Turnips, thoroughly well: and become very critical for Quality, & Quantity of the Harvest. . . .

Aug 2. Wrote Mr. Clutton, Tyringham to say I would take Mr. Lucas for a year, or more: as Agl. pupil.

Rent Audit – Return £350, Gratuity £250 – on current year!! £1200 for whole year – upwards of £300 expended in New Walls, Gates, & Iron Fence: in addition!!? always best to live under a 'large hold'.[1]

Aug 3. Gleamy peeps of Sunshine – Thundery atmosphere – not one really fine whole dry day in week.

Mr. Richd. Eagle came yesterday, & bot. 19 yearlings & calf £256.10.0d. – Finished thrashing last year's Harvest!!!

Aug 7. Jane & I, by 8.15 train, via Didcot, to Bristol at 12.45. Saw 40 Pigeons started from Didcot, to race Home to Birmingham.

Mr. E. Bullock, 16 Charlotte Street, Brandon Hill – drove us out to Penmaer Point, Durdham Downs, to Avon &c – had very fine views over and around the Bristol Channel – Dundry Church, very conspicuous.

Aug 8. Up at 6 a.m. walked around Brandon Hill – then down to river and saw Timber unloaded out of 'Nedjed' from Russia

By rail to 'New Passage', a rough voyage in Packet to Portskewett, and on to Newport & Cardiff, where we attended the 'Eisteddfod'. Brindley Richards played a Solo on the Pianoforte, grandly!!! he was the Composer of 'God Bless the Prince of Wales'.

The Marquis of Bute's Flag was flying on Cardiff Castle, which is a fine piece of Architecture.

Rain fell most of the day – before leaving we walked round some of the Docks – saw Coal Trucks of 10 Tons, emptied into 3 Steamers simultaneously by Hydraulic power – very cheap, expeditious. After rough rainy journey, & Voyage, reached Bristol 10 p.m.

Aug 9. By tramcar to Clifton Bridge, walked up the Zig Zag to Top, & across it, pretty views from it, in all directions – the Railway runs along the side of Avon to Portishead.

Mr. Bullock drove us to Leigh Court in p.m. – and we saw Sir Philip Miles' Paintings – by some of the *Best* Artists in History – one of them was said to be worth £40,000, the 'Creator'. 'Susannah & the Elders', was very good.

The late Sir Wm. Miles formerly shot over Earl Scarborough's Linc. Estates at Saxby, Spittal, &c.

Sir Greville Smyth has a Park & Mansion, opposite Miles' Estate.[2]

Witnessed the 'Merry Duchess', in evening.

Aug 10. Left Bristol at 12.15 for Bath, we looked round the Cathedral or Abbey, Bath Room & Pump Room, Park, and principal streets and places till 5.50 when we left for Charlbury at 8.39 – after a very variable visit, amongst all kinds of scenery – it is very pretty, & *picturesque* by Rail, from Chippenham to Bristol.

Mr. Bullock spent 2 hours in a.m. showing me round Bristol – pleased with the Council Chamber, & likenesses of various Mayors of Bristol.

Bristol is built on the Avon & Frome Rivers – away from the Severn, which flows from Gloucester into the capacious Bristol Channel – altogether I found the place, situated very differently, to what I had expected.

Aug 13. Shipton Auction – Mr. Price and I, won 9s. of Huckvale & Sargent at Whist.

Daylesford sold to Baring Young.

Duke Marlborough turned Conservative – was merely Radical to Displease his Family – who despised him for his alliance with the Marchioness of Aylesford, who they say has recently gone away with some other Noble Lord.

Aug 16. Mr. C.D. Beauchamp, overlooked the New Walls, Iron Fence, &c – allowed all – and took another Contract for 100 to 120 chains more, & ordered Mr. East, to send him a Contract, to force the Water out of Shipton valley, to my House, Yard, & 3 Cottages.

Aug 20. Began Harvest by cutting Little Brazil Wheat with Reaper: 1 Fagger in Hall's Hill.

Finished Turnip skerrying – carried 17a. 2nd crop Cow grass out of Farthings – & all the Lambs on Turnips.

Aug 22. Yesterday I accd. Mr. Price & John Rowland to Cheltenham, where we witnessed a Cricket match, between Surrey & Gloucestershire. The 2 Graces, Moberley, Gilbert, Townsend, &c: played grand Cricket, but were beaten by Surrey.

Aug 31. 9 waggons carrying wheat – rain at 5.30 p.m. 7 wheat Ricks up – 40a. Oats carried – & 13a. wheat in Covered Shed for Seed – had grand weather the last fortnight.

Sept 7. Mr. Jas. Fras. Mason, Eynsham Hall, inviting 100 Farmers, & others, to a Ball at Witney in Oct – in comn. of his coming of age, last year – deferred thro' a Sister's Death.

Sept 9. Magnificent warm sunny morning. At Church in p.m. Marwood, the public Hangman, died last Tuesday in his 64th year, he resided at Horncastle, Linc. – practised hanging 12 years, & used the Long Drop.
Mr. Price & I walked thro' Big Field, & admired its good crop of Oats.

Sept 10. Rain set in at 7.30 p.m. – & much fell the next 24 hours – to the great dismay of Farmers, & injury to the grain – my 90a. Barley ready to commence carrying out of Big Brazil – hoped to have my very best week of Harvest done – alas – 'Muck & Water' – and this has been our Fate for years – to the ruin of *Thousands* who cultivate the soil of Cloudy, Foggy, Blighty, Rainy, *uncertain*, Old England!!!

Sept 13. John Chas. at Witney arranging with Messrs. Bateman, Gillett, Bryan, Shillingford, & Jas. Fras. Mason: about his Ball & entertainment – to his Tenantry, & Hunting Friends.
Mr. John Addy's Self-binder has cut about 11 to 12 acres per day.[3]

Sept 17. Splendid day – carried Oats till 10.30 & then 21a. loose Barley out of Shipton Field – Dores & 15 women, heaped 32a. Barley – & turned 5 acres. . . .

Sept 20. Rain at 8 a.m.
Drove to Witney & tea'd with Mr. Wm. Worley.
Mr. Abraham had his Jaw broken, & otherwise injured, by their Elevator crank.
Mr. Wm. Shillingford, & Mr. Flood, beat Mr. Wm. Worley, at Billiards.

Sept 24. Most annoying weather: rained from Midnight till 9 a.m. & then 4 hours drizzle – with much Corn to cut at Shipton – & still more to carry all around – the 2 Hobbises, all their Barley out, but 4 acres – in fact daily spoiling all the latter corn.

Sept 26. An Inspector called, and arranged for a Pillar Post, to be built in wall, in Farthings spinney by the 'Corner', which was a great meeting point, in the Old Forest days. . . .

Sept 29. Showery, carried no Corn, since 19th ult – *Thousands* acres between here, & Cheltenham, spoiling in the Fields.

Mr. Tom Eve wrote to say Beast were dying in Linc of 'Foot & Mouth'. Weather spoiling all the latter Harvest – never saw so much *bad conditioned* grain, offered at Louth: Grazing Season ruined, by restrictions.

Can only Plough, Thatch, & Fag Beans!!!

Sept 30. . . . This month has Half of it, been too wet for Cutting, or carrying, White Corn: but very growing for Grass – Clovers – & Turnips: and free from Frosts.

All my Carters, Foreman, Shepherd, Groom, & Labourers, stay on, at about the same wages – except James Pratley who raises 1s. week.

Oct 16. Accd. John Chas. & Elizh. to a Ball in Witney Corn Exchange, given by Mr. Mason, Eynsham Hall, to all, whose land he has Hunted: with his Harriers, &c – in remembrance of his Coming of Age – since which date his Father has allowed him £10,000 a year.

Played Whist with Mr. Jeffery, Blk. Bourton: with Mr. Townsend, Witney: & Mr. Eagle, Stanton Harcourt, as adversaries – won both games.

Oct 25. Drove to Witney – Having my Hunting clothes prepared for the Winter's Campaign amongst the Reynards.

Dull, with high wind, & slight rain. . . .

Oct 30. . . . Finished Harvest – all Beans out of Boynal Gardens, 5 Ricks, the greatest crop, I ever grew.

Oct 31. Saw Mr. East, Burford: told me Caudal Spring water was to be forced up to my House top – and supply House, 3 Cottages, all the Stables & yards, Shakernook, &c – a Grand improvement to my Occupation – and the *Best* item, I have ever had done. . . .

October has been a wet, foggy, hindering month, for finishing the Harvest – several Bean fields to carry yet – the Wheat goes in very *heavily* – 6 Horses, & *Press* Drill – Corn in *bad* condition, for thrashing. . . .

Frosts have kept off – and the Grass – young Clovers, & Turnips all most growing, & a verdant Green!!!

Beast all out but 38 weaned Calves.

Sheep all flitted from Leafield Turnip piece (17a.) to Cockshot-hill, next Forest end:–

Mr. Hobbis' Charlock, amongst Turnips, in *full* bloom like *June*, & quite as gay in *Colour*!!!

Nov 20. Completed, & finished all up, for the Winter – on behalf of Mr. Groves 1st & 2nd Walls contracts.

Lord Randolph Churchill Hunts a first class pack of Harriers in Style – he is a clever Horseman – and Politician!!!

Nov 21. Isabel's 7th Birthday – Money – Doll – Cards – Chocolate Creams – & other presents, graced the Breakfast table.

Mr. Chas Hamblin wrote me saying the Badminton Hounds had been out 68 times last Saturday, & killed 132 Foxes. . . .

Nov 25. . . . attended North Leigh Church – dined at Mr. Gask's – and to Ramsden Church in afternoon – pleased with the Revd. Baker's sermon – Mrs. Mason, & her 4 Daughters from Eynsham Hall, were at North Leigh Church, which is a very fine ancient Building, & contains the recumbent effigies of Lord & Lady Wilcote, besides other rather important Monuments – &c.

Nov 29. Mr. Groves placing gates at end of Brazil – 4 men Walling – 2 digging Stones, & hauling, & wheeling.

15 men digging trenches, & Water-Wheel House out, in Coldwell valley – 120 nine inch pipes in – & hauling Stone down Shakernook valley, for Wheel House, &c. . . .

Nov 30. . . . This has been a month of great Floods – have only drilled Wheat, 7 days during this month!!! Lincolnshire is in a fearful State of Flood, Lord Willoughby's land at Tothill & Gayton, *all* to re-let!!![4]

Dec 3. . . . Mr. W.R. Allenby, So. Thoresby (Lincs), partly cut his Throat – strange to say, Mr. Sleddle who lived same Farm & House, cut his Throat about 40 years ago.

Alice & Mr. Price accd. me, & we called Mr. Payne's, Finstock – & afterwards met Ld. Randolph Churchill and his Harriers, at Fawler – wild, mild, windy, boisterous day – Ran 4 or 5 Hares – but scent bad – altogether 18 of us, & 21 Harriers.

Dec 5. Met Hounds Ledwell – Luncheoned at late Mr. Eve's: Mrs. Caldicott there – found in Worton Heath, soon lost – frosty air.

Col. Thomas, said Ld. Randolph Churchill, would like a day with his Harriers, over my Farm, shall be glad to see him, after the Hounds next visit.

Dec 31. Met Hounds at Chapel House – did not find till we reached Salford Spinnies – ran a ring & lost – left for Home at 2.30 – Hounds went on to the Alcombs &c – been out 22 times!!! since the 20th Oct. ult.

Taking Agriculture *all round*, the past year has been the best I have experienced in Oxfordshire: and one of the very *worst* in Lincolnshire.

1884

Jan 1. Fine dry leaden looking morning.
Posted a letter to Mr. A. Brassey, & included a Brace of Hunters for his Sons to spin round in their Nursery!!!
Arch over Water-Wheel House finished by Mr. Norton, Chilson.
Tea'd & Bezique, with Mr. Price.

Jan 7. . . . All our Family, except Isabel, attended Witney Childrens' Ball – Mr. Ward accd. them: Alice, Eliz., John Chas. Robt. Wm. – Grace, & Fredc.

Jan 15. Met Lord Randolph Churchill, & his 24 Harriers, at Mr. Abraham Stratton's, Crawley. . . .
Dined at Churchill Arms, Ascott: with upwards of 50 members of the Shipton Conservative Association – John Scovell in Chair – Mr. Stapleton, Stamford: spoke the best –
Beautifully Dry, mild, still weather: most favourable for Agriculture.

Jan 21. Alice & I, drove to Chipping Norton, & Mr. Price drove us to Banbury Horse Show – looked round the Church &c – bot. 17 of Betts' *Real* Banbury Cakes: Home 6.45.

Jan 23. Grace with the 2 Miss Gasks to Miss Robinson's Seminary, Balham Court, London S.W.
Very high wind, with rain squalls: no work done. . . .
Miss Graham from nr. Carlisle, came as Governess.

Jan 24. Drove to Witney – Saw Mr. Robt. Abraham – who has just returned from Mentone – said he had brot. me a Present – as a memento: Mr. C.H. Spurgeon, he left behind, helpless with Lumbago.

Jan 30. Joseph Reade junr., Shipton; 21 years old *this* day: dined at 6 p.m. – about 250, of the Labouring Class.
Mr. Price gave me a Shooting Tub, for Fairspeăr Clump.

Jan 31. Had an extraordinarily mild Month: only 70a. to plough for Spring Corn. . . .
Paid Mrs. Morris 22s.3d. for Mr. A. Brassey: Reynard blamed for worrying 4 couples Fowls, 4 Hens, & 1 Duck!!!
A mild rain.

Feb 12. Spent the evening with Mr. Thos. Brooks, Shipton: paid Miss Brooks, 10s. for her Xmas 15 lbs. Goose.

Windy & mild – Country looks as fruitful as latter end of May!!!

Feb 19. Mr. Price called at 7.15 and we reached Mr. Rowland's Wootton at 9 – left at 10.15 and reached Bletchingdon Park House at 10.45. Visct. Valentia invited us to Breakfast – had Champagne &c – at 11.15 rode away to Busby's Spinney (Osier Beds beside the Cherwell) found & ran thro' Harrises Gorse, or Frog's Nest, over Kirtlington Park, away towards Middleton – turned and killed in the open, half a mile from Heyford Stan. at 12.30 – ran very *fast* – Fox *sat down* dead beat. Luncheon at Wootton, & reached Home at 5 after a very enjoyable day – rain set in at 4.45!!!

Only 31 Bitches composed the Pack.

Ld. Chesham, Honble. Ponsonby, 2 Simmonds, Smith, Jones, &c – were out.

Mr. East & Norton – fixing & building in, the new Waterwheel, nr. Coldwell Spring.

Feb 21. . . . Evening with Mr. Price – tasted Devonshire Butter, Stilton Cheese – & Thos. Ellises Bread, *all* first class. . . .

Feb 23. Hounds met at Ascott . . . Been out 5 days this week – for the first time in my Life: with Hounds . . . I came away at 2 p.m. after which frequent showers. . . .

Feb 29. Mr. Gask, & Son Frank, Tea'd, & looked at Water Wheel. I spent evening at Mr. Wm. Ellises – sold 20 Tegs at Auction 53s.10½d. each.

February, been a very fine Month, on the whole!!!

March 1. . . . The Cold-well Spring Water, forced by Wheel to top of House and Tank House this day.

March 12. Met Hounds Addlestrop 12 – left at 1. Cherry Brandy, & Gingerbread. Major Arnold very hospitable.

March 19. By 8.15 train via Oxford, to Banbury at 9.25, upwards of 40 Gentlemen & Farmers, met at the Cross at 12: Chasers to Bourton, the Seers to Shotteswell Hill – Mr. Geo. T. Drake 1st Ld. Willoughby 2nd Campbell 3rd of the 12 st. weights, & Capt. Becher 1st & Whitworth 2nd of the 14 st. weights and Mr. John Walter, Farmer, 1st of the 12 st. weights and Mr. J. Whites 1st of the 14 st. weights – a regular old fashioned Steeple Chase – Ralph Hambidge ran on 'Runnymede'.

March 29. Attended a Conservative meeting at 2 St. Aldate's, Oxford; Holford Risley in Chair – A.W. Hall – Fox – Baker – Morrell – Barnett – Lockwood – Addy – Brooks – Bulford, & over 30 others.

Been a dry cool week with 3 rime frosts, drilled Oats every day.

John Chas. & Alice, hunted at Dytchley, killed a Fox, and Badger.

March 30. . . . Mr. Price walked down to Wheel-house, & saw all thoroughly completed: the wall around, makes a good finish.

March 31. Tea'd at Mr. Price's after Shipton Auction – Rain set in at 12.

Been a fine mild Dry March: 120a. Beans, & Oats, drilled.

On 15th ult. Weston, finished his wonderful walk of 5000 miles in 99 days –!!! a *wonderful* performance, of endurance!!!

April 10. Last Season ending 6th April 1883, I met Hounds 32 times, this Season ending 9th April 1884 – I met Hounds, & Harriers, 52 times, being the two greatest Seasons of my life . . . A very good Season, Mr. Brassey's 11th year's Mastership – killed 77 Foxes, and a Badger in Dean Grove, & an Otter at Barrington.

The Duke of Beaufort's Hounds have killed 240 Foxes *this* Season – & 2 Badgers.

April 19. Sir Wm. Bazley re-let, one of his Farms, to a very old Tenant, to hold 3 years Rent-free, & Owner, to pay half the Labour – shows Agriculture depressed indeed.

April 24. Visct. Valentia dined with 35 of us, at Witney, was introduced by Mr. A.W. Hall – he comes forward at next Election in place of Col. North, who resigns, thro' Old Age.

Tea'd with Mr. Sutton, Blake, Smith, & Pratt at Marlbro'.

Elizh. & Grace, did shopping, & tea'd at Robt. Wm's.

Recd. a depressing letter from Bennett Farrow, on Farming, in East Lincolnshire.

April 26. *Finished* all Walls, Gates painting, & cleared up all the stones, from the Waterworks trenches, and completed everything, till after Harvest: ready for Mr. Clutton's inspection.

An Earthquake felt in & around Colchester – much damage done in some of the Villages.

April 29. Drove Jane to Northmoor, and attended Miss Rachel Eagle's Wedding: Revd. Tuckwell, Simmons, & Walsh, performed the Service – Mr. Henry Bulford, the Bridegroom, made a very good speech: all passed off very nicely – 4 Miss Bulfords, & the 4 Miss Eagles, were Bridesmaids.

April 30. Accd. Mr. Gask by Rail to Cheltenham – witnessed the Races, & 2 Steeplechases.

Been a Dry – and latterly, Cold Month.

Mr. Abbott wrote me from India. . . .

May 5. Grace retd. to Balham Court, after her *first* Holiday from School – she frequently thinks of 'Home Home Sweet Home'!!!

Mr. Jas. Shillingford called & tea'd, on his return from a bad Sale of 1000 Sheep, at the Junction.

Thunder Showers.

May 8. Waggonette tie broke, a mile off Witney. Attended an Agricultural Lecture, after which I tea'd with Mr. Sutton, Wilcote: at the Marlborough – & we were the 2 *last* to leave: 8.15 p.m. only.

May 10. . . . John Chas. ill, of the 'Mumps'.

Isabel, told her brother Fredc.: Currants & Raisins, were *imported* from Witney!!!

Sowing small seeds, shearing sheep, Dragging & ploughing Fallows, 3rd time for early Turnips: & men weeding Fernhills, thistley wheat.

May 12. While the Queen was at Darmstadt[1] – she, her eldest Daughter the Princess Fredk. Wm. – her Daughter – & Grand-daughter: were all 4 generations, Photo'd together – a short time ago, the Emperor of Germany – his Son – Grandson – & Gt. Grandson, were all 4 taken together – so the 4 Emperors, & 4 Queens, will be great *Companion* pictures in *Future* years!!!

The *'Victorian Era'*, will leave a *grand mark*, in History. . . .

May 16. . . . By 8.29 to Oxford with Jane, & Alice, we witnessed the Boat races at 4.30 & 6.0 p.m. – and a grand Cricket Match between the Australians, & Oxford University with the following result, of the 3 days performance –

Australians 148 & 168

University 209 & 110 & 7 wickets to fall!!!

A : 11	U : 11
P.G. McDonnell	M.C. Kemp
A.C. Bannerman	T.R. Hine-Haycock
W.L. Murdoch	A.R. Cobb
G. Giffen	J.H. Brain
G.J. Bonnor	T.C. O'Brien
J. McBlackham	H.V. Page
W. Midwinter	A.G.G. Asher
H.J.H. Scott	K.J. Key
G.E. Palmer	B.E. Nicholls
H.F. Boyle	H.O. Whitby
F.R. Spofforth	E.W. Bastard

. . . On 23rd the Marylebone C.C. at Lords' Ground, beat the Australians by 115 runs in *one innings* of 481 – A.G. Steel 134 – W.G.Grace 101, and Barnes 105, & *not out* at *finish*. *Unprecedented Cricket* against such able performers!!!

May 17. The Honble. Alexr. Gore, & Mr. John Clutton, as Comr. & Receiver for the Crown – Luncheoned, & spent 2 hours on my Farm, examining the Water-works – new Walls – &c. . . .

May 18. Magnificent day – Country looks charming – Tulips in full bloom on *North* side of House – Cherries & Plums, 'setting'.

At Church in p.m. with Mr. Price.

Cart Horses out, for 1st time, in Shakernook.

May 19. Rode to Fawler Iron and Clay Works, saw Mr. Collis, who walked with me over the Works, and under the ground to iron ore.

Rabbits bred early – thousands of young ones in the Forest nearly half grown – most *suitable* for Phts. & Partridges laying: every prospect at present of a good Game year.

May 20. Tea'd with Mr. Reade – and attended a meeting at Red Horse & formed a Limited Liability Co. to build a Club House at Shipton – for Political, & general purposes.

May 22. . . . Robt. Wm. retd. from Shillingford's & Co. Brewery – ready for another situation.

May 24. Queen Victoria 65 – hope she may reign over us 50 years, whatever more: what a mark her reign will be in English History – for Steam, on Land & Water – Telegraphs – & immensely increased Traffic all over the World!!! her 'Family' will rule over half the Universe.

Had a *very fine*, dry, sunny week.

May 27. Mr. Reade's Rent Audit at Shipton.

Rode with Mr. Price to Mr. John Dutton Blisses Farm, nr. Minster, and met Mr. Blandy Jenkins & his Otter Hounds – found brace Otters in a hole under a Willow Tree against Minster Weir – Hounds worried one, & Mr. Knight, Witney – killed the other – same time – tried a mile up beyond Minster Mill – Hounds gave Tongue & ran along river & out into a Drain nr. a Mr. Clack's House but could not get at the Otter – so ended our Sport, I dined at Mr. Abraham's with Mr. Mrs. George – May Palmer, Mr. Price & Mast. J. & Miss Madge Abraham.

May 28. Cool drying East wind – Rain much required – to make the ground warmer, & more growing.

Mr. Wm. Ellis, Shipton: very ill of Jaundice &c.

Attended a meeting at . . . Red Horse Inn, Shipton, to arrange about Mr. Hitchman's land, & writing the Gentry: for Capital with which to build a Hall for political & other Business. . . .

May 30. . . . At 2 a.m. on Thursday Wm. Ellis, Shipton: died ae 53¾ very much respected by a multitude of acquaintances, who had had business transactions with him, for many years!!! . . .

The Revd. H. Ward B.A. went with Alice by Rail to Oxford, and took his M.A. degree. . . .

May 31. Been a very dry May – never saw so much dust on the highways for years – as now.

Drove Jane to Mr. Reade's, Shipton: walked round the Swannery, & Court House, & Gardens.

I joined Mr. Price, & we attended Mr. Wm. Ellis' Funeral – he has no near relatives – 2 Cousins, & several neighbours, & Strangers were there – Mr. Mace, Solr., & Mr. Haigh, Surgeon: were there.

Horse 'Gilbert' out to grass at 7 p.m. – found groaning & straining, at 11 – in Shakernook: died at 1 a.m. of Dropsy.

June 7. Heavy showers again last night, & during the day – only Shepherd, & 5 others on the Farm – Leafield 2nd day's Feast.

Fredc. George & Isabel, walked up to Leafield Fair in p.m. with me – we bought Nuts, Gingerbread, & Rock!!!

June 11. Attended a meeting at the Red Horse: £160 already subscribed towards a 'Beaconsfield Hall' at Shipton.

June 13. My 30th marriage Anniversary – what changes have occurred to myself, & all my acquaintances of 1854!!! . . .

Revd. Carpenter & party – playing a scratch team, at Cricket, from Oxford, on Ranger's Lawn.

June 15. Mr. Haigh's servant maid, found dead in Bed, last Thursday morning, from Apoplexy.

At Church in p.m.

June 19. . . . Mr. Jeffrey's 3 Carters all dead of Typhoid fever.

June 26. Began carrying Clover with 9 waggons: out of Big Field 48a. & 30a. of Shipton Field – splendid condition: double hands.

June 27. Magnificent hot, close, sunny weather.

Fredc. Geo: Delirious in night, from Disordered Stomach – 2 pills were a cure: for which he asked!!!

June 29. During early morning, and the day, a succession of *grand* Showers – which were so very much required to carry on the Turnips – & in fact *all* other Crops!!!

Have Ducks, new potatoes, Cherries, & Strawberries, all ready for the

Table – Peas require another week – which shows the great difference a Sunny Season makes to all vegetation &c.

July 1. Trimming Trees in Shrubbery according to Mr. Clutton's orders – to the great disgust of my Wife & Family, who do not like to cut anything down.

July 31. . . . Beautiful weather for everything!!!

July, been a showery, growing month: Corn & green Crops done well.

Aug 7. . . . Yesterday carried 2nd crop Clover, Shipton Field, in *good* Condition!!!

All work nicely finished up for Harvest!!!

Morello cherries a good crop.

Turnips *all* skerried, & doing nicely.

Aug 8. All hands commenced tie-ing Wheat, in Kingswood, & reapers cutting Wheat, in Farthings.

Mr., Mrs., & Miss Batchelor, & Miss Anne Robinson, drove up from Milton, & spent the Evening – Mr. B., drove me round the Farm, & walked to Waterwheel – they have come to spend a month at Milton, instead of visiting the sea – Croydon is his Home.

Aug 12. The 5 eldest, with Misses Bullock & Scudamore, were driven to Witney Horticultural Show –

John Chas. took Owl prize.

Attended meeting at Red Horse – A. Brassey, A.W. Hall, J. Reade, Dr. Haigh, Mr. Addy, & myself: were appointed Directors, of Shipton Conservative Hall.

Supped with Mr. Reade.

Grace Catherine 16 years old.

Aug 15. Slow trade for the Rams at Oxford, great sales on Wednesday: Stock will be lower *all* round till Spring – and then the Trade will be ruled by the Summer-grazing Prospects.

Aug 18. Earl Ducie cannot 'let' his land – advertising weekly!!!!

The Crops of this locality, are Whiter, & Brighter, than we have had for years, & Corn *good quality*. . . .

Eliz. gave her Sunday School Form, Tea &c on the Lawn, a delightful evening for them.

Finished 275a. Wheat cutting, & cut 50a. Oats, already!!! . . .

Aug 19. From Shipton by 9.34 train to Cheltenham – looked round the shops, Parade, &c – to College Cricket ground at noon, to see the Australians play against Gloucestershire. . . .

Fairspear House, 1895. (By kind permission of the executors of Mrs. D.E. Hall.)

The Calvertt Family outside Fairspear House, 1884. From left to right:
back row: Elizabeth Anne, John Simpson, Jane, Robert William, Alice, Grace;
front row: Isabel, Frederick George, John Charles.
(By kind permission of the executors of Mrs. D.E. Hall.)

Aug 20. . . . This day twelvemonths began Harvest – tonight, cut 392a. Wheats & Oats – carried 4 days to Machine, and 5 Wheat Ricks up –!!! Grand Weather!!!

Aug 28. Drove Alice, to Witney, with Isabel –
White Wheat, was sold at 34s. and Red at 31s. & 32s. per qr. – the *cheapest corn market*, I ever attended, in my Life!!!
Bot. a £3 Game Certificate: paid for 37 since 1847!!! missed Sept 1852!!!

Aug 31. Frequent rains during the day: had a very showery walk in aftn.
Augt. has been one of the grandest month's weather of the year – during which, all the early Harvest, is secured, in good Condition!!!

Sept 6. Drizly rain at 11 a.m. – Sent all Home out of Harvest field.
Mr. Norton turned *all* the Coldwell Spring Water, on to the Wheel – which works well – and will prove a *never failing* supply!!! the 'Spring Head', is exactly 10 yards from Mr. Hambidge's dead hedge!!?

Sept 16. The 12 Australian Cricketers, named in this book, on 16th May last, began 32 matches on 12th May ult, & finished on 11th Sept ult – they won 18 – Lost 7, & 7 were Drawn!!!
Met Mr. Reade & Thos. Brooks, Crown Inn, Shipton: and we arranged to take Mr. Groves' 'Contract', for Beaconsfield Hall, subject to alterations!!!

Sept 20. Finished carrying Harvest – in Grand Weather – 513a., all cut with 2 Reapers except 6a. 3r. 0p.!!!

Sept 24. Hired to stay again, Foreman, Shepherd, Groom, 3 married Carters, & 3 Labourers – all same wage as last year.

Oct 10. Elizh. Anne daily packing up for Bombay – she sails on 22nd inst: has received several nice presents.[2] . . .

Oct 19. Elizh. Anne attended her Sunday School Class, and told the children, she was about to sail 6000 miles over Water, to Bombay!!!

Oct 20. By 11.28 train from Charlbury, to Paddington, from Praed St. to Aldgate, walked to Fenchurch St. Stan. & away to Central Stan. Royal Albert Docks – saw Capt. Turner & over the S.S. 'Kerbela' – retd. to Salisbury Hotel for Tea – and saw the 'World' performed at Drury Lane.

Oct 21. From Blackfriars Stan. to Victoria – met Grace from Balham & retd. for Breakfast to Salisbury Hotel with Jane. Elizh. & Wm. spent all day at the Healtheries Exhibition, adjoining Albert Hall: very pleased with all we saw, and Dan Godfrey's Music!!! Saw 'Called Back' at the Prince's Theatre, Coventry St.

Oct 22. Walked to St. Paul's before Breakfast – bot. an Umbrella – Jane, Elizh., Grace, Robt. Wm., & myself left the 'Salisbury' in a Cab for Fenchurch St. & by rail to Royal Albert Docks, were met there by Mr. Price, Ward, & Alice – looked over the 'Kerbela' – saw the passengers for Bombay &c – the Louise & Beatrice Tugs – hauled them off at 1 p.m. in delightful weather!!! we journeyed to Westminster Aquarium very amused with the Circus performers, and the two Brothers, & Sister Beckwith's swimming!!! left Paddington 6.30 – Charlbury 8.40.

Oct 25. Recd. a letter from Elizh. Anne off the Isle of Wight – 25 First, and 12 2nd Class passengers in the Kerbela – 3 clergymen, & their wives, amongst the party: weather delightful!!! . . .

Oct 31. October has been a very fine, mild, Dry month!!! quite a water famine, in many Counties. . . .

Nov 2. Rode over to Mr. Gask's, Bridewell: found him laid up in Bed, after being kicked over a Gate by his Horse, and sprained the muscles of his Neck &c –

Rain at 2 p.m. to my great satisfaction.

Nov 3. Rode and met the Old Berkshire Hounds at Yelford . . . Povey is the new Huntsman, & was whip to Dan Berkshire in the Southwold!!! so he knew me at once!!! . . . Called & had lunch at Mr. Adams, Hailey Manor . . . Tea'd with Mr. Price, who had attended Heythrop opening!!

Nov 13. 55 years old – Mr. Abraham & Toller tea'd.

Isabel presented me with a Tie –

Wheat sold in Mark Lane on Monday last at 23s. per qr.[3]

Foreign Flour 17s. to 21s. per sack of 20 stones. Potatoes 2s. 3s. & 4s. per 12 stones, or sack – Pork 4d. & 5d. lb.

Selling my own Blk. Oats 17s.3d. 17s.6d. & 18s. per qr. 40 lbs. Bushel!!!

Nov 26. Wrote Lizzie, a letter – the first she will receive from us, in Bombay: 5d. stamp for ½ an ounce.

Nov 30. Drove to Northmoor, & dined with Mr. Mrs. Richd. & Blake Eagles and stayed night – snow, & then rain set in.

This has been a fine, mild, Dry month: similar to the *10* previous ones.

Dec 5. Rain more or less *all* the day – mild & still – and previous night, with wind: so expect the Tanks, & Ponds, will be supplied with water for months – & the cleaned & cleared out Ponds, for some *years* – never known water, so scarce, the last 9 years, in Oxon.[4]

Dec 11. John Chas. drove his Mother to Witney – stayed a performance in the Corn Exchange, given by the 'Eynsham Party'!!

Dec 14. 23 years today since the Death of the lamented Prince Consort.

Mr. Robt. Price walked up, and Dined, had Mr. Brassey's brace Phts. for Dinner – drank to the Queen – and Mr. Mrs. Brassey's good health, hoping for many years to come, they might have as good a Show of Foxes & Pheasants, as this Season!!! Shot 400 Pheasants, & have abundance Foxes!!! . . .

Dec 20. . . . Walked down to Shipton Post Office – and into Church yard, looked at Wm. Brooks' grave, only last Monday, he walked from Ing's Grave till we met Mr. Price, with whom he had been Foreman &c – for 30 years – he died suddenly of Heart Disease, 9 to 10 p.m. on Wednesday night, ae 61 – Honest, industrious man: his wife died 9 months ago, his only Son & child, was married 1st last January.

John Brooks, brother of the above, has been a Policeman in Oxford, 24 years.

Wind, Snow, & Rain – prevented me, meeting the Hounds at Blenheim. . . .

Dec 25. While at Church a.m. – the Postman, left the 'Times of India' in which we found Elizh. Anne's marriage, to Algernon Abbott, G.J.P. Railway: at Christ Church, Bycalla, Bombay – May Good Health, & Good Luck, attend them.

Mr. Price came & tasted Pineapple, before Church, in p.m.

Fine clear day, with slight Frost – and most things as green, & gay, as Summer – Primroses, in Covers!!!

Dec 26. A very interesting letter from Mrs. Abbott – with particulars of her marriage – Presents – Voyage – Capt. Turner – Her New Home – altogether she has received a very good impression of Bombay, and its Climate.

Mr. Herbert Ward, came yesterday, for 2 nights.

Dec 31. Met Hounds at Junction – found at Gawcombe – & lost towards Bourton on the Water – found 2nd Wick – left them in Gawcombe – too frosty cold & hard – very little scent.

This has been one of the best years for *suitable* Weather, for *Farmwork*, on record.

Good Crops, & good Quality – & cheaper than I ever grew them, all round.

1885

Jan 5. Miss Pearson came as Governess to Fredc. Geo. – & Isabel.
Met Hounds at Chapel House – tried Walk Gorse, Salford Gorse & Osier Beds, Over-Norton Park, & Adlestrop Hill – *Blank* – came away in a drizly Fog!!! rather dissatisfied with such useless waste of Time, in what ought to be, the prime of the Season.

Jan 15. . . . Earl of Aylesford dead ae 36 – so the Duke Marlbro' can marry the widow, who was a Miss Piers Williams, of Bangor.

Jan 20. Grace by 8.25 train, to Mrs. Dix, Merton College, West Croyden, John Chas. accd. her.
Alice, to Oxford – attended a Grand Concert at which the Princess Christian played duets on the Piano, with Mr. Parratt, organist, St. George's, Windsor: the Sheldonian Theatre was crowded – & consequently the Concert a great success. . . .

Feb 12. Alice & John Chas. to Witney.
I dined with C. Chas. Horne at Crown, Shipton.
General Gordon killed in Khartoum: thro' Treachery!!![1]

Feb 16. Fog and Rain, all day – Arranging for Alice's Wedding, on March 12th.
The breadth of Wheat harvested in America was 39,475,000 acres in 1884: the *average* yield *one* of the *best* known, *13 Bushels* per acre!!! worth on the spot 1s.6d. to 2s.6d. per bushel!!!
The United States, Russia, & India, are our greatest importers, we are the only Country, who require *always* to import a large proportion of our Consumption of Wheat.

Feb 22. Ground covered with Snow 6 a.m. Rain set in at 8 a.m.
Jane to Church to witness Alice's 'Bands' the 3rd time.

Jan 27. Rode down to Shipton Sale – sold 48 sheep. . . .
Hounds met at Tangley – Mr. Brassey laid up of a bruised face and black eye – from his Horse slipping back over a drop jump, and suddenly throwing his head back.

March 4. Alice, has a Case of Silver given her, of 84 pieces, viz 18 Forks, 12 Table spoons 12 Dessert spoons 12 small Forks 8 egg spoons, 4 Salt spoons, Mustard spoon, Sugar spoon, Sugar Bows, Butter knife, Soup Ladle, & Gravy Spoon, & 12 Tea-spoons. Also Silver Kettle & Tea & Coffee Service – Biscuit Box, 6 Silver Salts – Toast Rack – Fish Carvers & Knives & Forks – Gold Brooch – Lamp – Sugar Scuttle – Chronometer – Watch – Brass Candlesticks – China Tea Services – Cruet Stand – Communion Service & Vestments, 2 pairs Oak & Silver Carvers – Silver Pin – Revolver – Vases – Album – Gold Brooch – Scott's Poems – Dessert Knives & forks in Silver – Travelling Clock. Cheques for £50 £21 £20 £20 £10 £5 Sardine Box – Silver Marmalade Pot, Jam Dish – Nut Crackers – Gloves – Time piece – Pincushion – Fancy Box – Carriage Rug – Ivory Prayer Book – and altogether upwards of 80 Wedding presents.

March 5. By 8 a.m. train from Witney to Birmingham at 10.33 for 7s.1d. – Champagned at Mr. Matthews – attended Messrs. Lythall's Bull Sale at Bingley Hall, bot. No. 266 – 21 months old – 23 Guineas – 'Baron' by Duke of Warwick 6th 46274 Dam 'Bride' 33576 by Duke of Barrington 6th 49041. The Bull and I reached Shipton Stan. at 8.12 p.m., 4s.11d. Self & Bull 23s.9d. in a Horse Box. Lunch at Messrs. Sutton's Stand.

John Chas. attended 'Moreton Steeplechases' – wined &c – with Messrs. Henry & Albert Brasseys – Midgeley – &c.

March 11. Fine cool dry morning!!!

Alice Louisa Jane married to Herbert Marlow Ward at 11.30 at Leafield – by Revd. J.W. Lee, assisted by Revd. H.P. Ward, Bridgnorth, father of the Bridegroom – 6 Bridesmaids – Grace Catherine Calvertt, Emily Calvertt,[2] Isabel Calvertt, Miss Clara Oliver – Miss Edith Westell, & Mr. Bottomley the Best Man – Jane, John Chas. & Robt. Wm. with R. Price, D. Westell – &c. – went to Church – a great crowd of Witnesses – Dr. & Mrs. Haigh – Mrs. Gask – Mrs. Lee – Mr. Wm. & Miss M. Abrahams – Miss Savidge, Miss Timms, & a host of others.

Bride & Groom, to London, by the 4.38 train from Ascott.

Miss Nellie Bullock had to keep to her Bed, and C.D. Batt, pronounced Mild Measles!!! to the great regret of all.

The Revd. H.P. & Mrs. Ward, returned to Bridgnorth from the Junction, direct for Worcester, for Bridgnorth.

March 24. Mr. A. Groves found good wall stone, in Leafield piece, to my great surprise – the substrata varies wonderfully all over Fairspear Farm.

Rent fixed from last Michs., to Michs. 1888, at £1155.[3]

March 26. . . . Mr. Price drove Revd. Ward, Mrs. Ward, Miss Oliver, Grace & myself to the Junction, Mr. Brassey arrd. there at 1 p.m. – and ordered a large concourse of People to Idbury Top – to witness 4 Point to Point Steeplechases – John Charles rode Mr. Price's Mare 'Windsor' and easily won the 2nd 12 st. Chase – to our great gratification!!!

April 18. . . . The Revd. H.M. Ward, & Alice . . . sail for Belize, British Honduras, on 8th May. . . .

April 25. Had 3 windy days, with Showers from South West – but cold. The Blackthorn in full flower in Forest – rode thro' the 'Slates Copse', cut & cleared this Winter – as I ride up & down the 1200 acres remnant of Whichwood Forest, all in a piece, & *no public* road whatever thro' it – I think of its History, the Hunting Ground of the Kings of England for Ages, and what sport of various kinds it afforded them, nearly all the year.

May 1. Fine brisk morning – rain at 10 fine at 12 – rain at 3 with intervals of sunshine – twice wet thro'!!!

Mr. C.D. Beauchamp called, and examined the House roof – will certainly have to be slated.

Alice packing up for Belize – expects to start from West India Docks, on 7 or 8th inst. . . .

May 2. By 1.15 train to Oxford, met Mr. Holford Risley, at Conservative Club – Mr. Wynne, A. Brassey, Capt. Waller, Saunders, Reade, & 25 others – elected a Chairman, Vice-Chairman, Honble. Secy. &c –

Had wine with Mr. Blake – & also 'liquored up' with Mr. C. Symonds, nr. the Randolph.

Best Red Wheat 40s. – 11 st. Oats 23s. War expected, with Russia.[4]

May 3. At Church & Sact., with Alice – for the last time for some years, if evermore at all!!! . . .

May 4. Jane's Birthday –

Alice, left Fairspear, with her husband, the Revd. Herbert Marlow Ward, for London, till next Monday or Tuesday, on which day they both sail in the S.S. Salerno, for Belize, British Honduras – 1.5 train.

May 8. . . . Accompanied Fred & Isabel – with Miss Pearson – to witness Conjuring, & Mesmerism, by Roberto, & Astarte, in Leafield School Room.

May 11. Jane, Wm., Mr. Price & myself by 8.15 train to Paddington 10.25. Saw Mr. Mrs. Brassey & Mr. Adams – and were met by Mr. Mrs. Ward, & accd. them to 'Inventions' – by rail to Mark Lane, walked across to Fenchurch St. and by rail to West India Docks, saw the 'Salerno' 240 feet long, looked all over her, and returned to St. Paul's Crypt, Nelson is buried exactly under the Centre of the Dome, and the Duke of Wellington, 20 yards further South –

Saw Mrs. Langtry in 'Peril', at the Prince's Theatre.

May 13. Jane, Wm., Grace, Mr. Price, & myself met the Revd. & Mrs. Ward at Mme. Tussaud's – the first visit to the new Establishment, a very interesting place, especially for studying English History.

Grace bid Alice 'Goodbye', & left for Mrs. Dix, Merton Coll., W. Croydon.

Jane & I, bid the Revd. & Mrs. Ward 'adieu' at the Paddington Stan. and left at 12.15 Ascott 5.5.

May 18. Recd. a letter from Alice, written 15 miles E of Isle of Wight on their way in 'Salerno' for Dartmouth, Nassau, & Belize.

Signed Mr. Henry Childe's 'Will'.

Attended a meeting at Red Horse, Shipton: Messrs. Reade, Haigh, Brooks, Halford, & Young there – arranged that Mr. Groves Builds the 'Beaconsfield Hall' for £300 hauling partly excluded – &c.

Supped at Mr. Reade's.

May 19. John Chas. packing up – and arranging for a *long Journey*!!!

Mr. Price came in the evening to bid his Steeple-chase Rider adieu!!! . . .

May 21. John Chas. & Childe expected to start this morning at 10 – in the S.S. 'Sardinian', belonging to the Alan Line, would call at Londonderry, and then sail direct for the St. Lawrence & down to Quebec, and by rail to Toronto, which is situated on Lake Ontario – 'God speed them well'. . . .

May 23. . . . Signed 30 Beaconsfield Hall, Share Papers, as Director: and have taken Five, £1 shares.

June 1. Cherries setting – Magnums big as Peas – the latest Apples in full bloom – Acacia just shows signs of life – the 'May Bloom' not out: Bluebells, Tulips, and bed of Easter White Lillies, all in full flower, in *front* of House.

Fine sunny day, with cool breeze of North West – hoeing Beans.

All the Corn, looks healthy, and a full plant – a deal of the Wheats are extra big, & forward.

'Baron', & 'Fairspear', to Cows & Heifers: 57.

John Chas. reached Quebec, in the 'Sardinian' – this day.

June 9. Earth Stoppers' Dinner at Charlbury.

Very dull threatening day, after *12 hours* gentle rain.

. . . Government last night, defeated, on Sir M.H. Beach's Resolution,

For second reading of Budget	252
Against	264
Majority against Government	12

Amidst great uproar, Mr. Gladstone adjourned the House – am very glad to have a Change – quite tired of the present Ministry – Vacillation – has been their Down-fall!!!

June 11. Drove Robt. Wm. to Witney, called upon & arranged with Mr. Coles to accompany Robt. Wm. to meet Mr. Warn, Tetbury: at Cirencester on June 22nd[5]. . . .

Paid £3.16.6d. Sch[edule] B upon £153!!!!

June 15. The Marq. Salisbury, is about to form a Conservative Ministry[6]. . . .

Sent Mrs. Reade a black kitten – retd. a Melon.

June 22. Rode over, and spent evening, with Mr. Adams, Hailey. Mr. Dodson, Chase Farm, sadly backward with work, the wretched place will ruin him – belongs Duke Marlbro'.

Half my Turnips in – nice Pastures all round – Sheep and Beast *all* doing well – not *too* hot for them.

June 29. Splendid weather, for Hay & Clover carrying, & the Wheats.

Rode to Forest plain after Tea – met with F. Shayler, amongst 500 young Pheasants – told me they had 2 litters Cubs – and that Victor Spencer, would be 21, the 23rd Oct. next – the only child of Lord Churchill.

July 3. Drove Mr. Price to Chipg. Norton Kennels, and met Ld. Chesham, the Honble. A. Rushout, & Povey: as Judges of the 41 Puppies: Mr. Brassey, Godman, Thursby, Lockwood, Evans, and about 40 others came to Luncheon, &c.

Home at 7.30 by Sarsden.

July 7. Received a letter from Alice, posted at Nassau, one of the Bahama Isles – and two letters from Bombay – from Mr. & Mrs. Abbott.

Fredc. Geo. walked down with me, and admired Mr. Price's Roses and tasted the Strawberries.

July 11. Too dry for Turnips growing – a good rain is required for nearly everything. . . .

July 13. Shipton Auction – 630 Sheep – 35 Beast.

Received a letter from Alice, written in Belize, describes the Island of Nassau – they abound in Fruits, Flowers, Sharks, Lizards, Land Crabs, & Sand Flies: but have an abundance of Turtle Steaks, & Soup!!!

The two Brothers Parsloe, began stripping & Slating, the House: which will be a great improvement.

July 21. Mr. L.M. Wynne, came, and addressed the Leafield Electors – probably only the 2nd time a Political meeting has ever taken place, since the place was a Parish.

July 29. John Chas. reached Home from America – after visiting Quebec – Montreal, Toronto, Niagara Falls, Philadelphia, & New York. . . .

July 30. . . . Richd. Eagle junr. threw over Mr. Whaley & himself out of a Trap nr. his own House, Northmoor: broke his own head, & 2 of Mr. W's ribs – had been to Mr. Brassey's Ram Sale.

July 31. Wheat & Barley – Beans & Peas – cut in several localities.
The past month has been very Dry and Hot – no Turnips, much of the Spring Corn prematurely ripening, and scarcely any trade for Sheep, and Beast: Fodder, & Wheat Crops, will be average – all the rest Deficient. Wool 18s. to 20s.
Most favourable season for cleaning Arable land.

Aug 4. Mrs. Sartoris 41 'Engaged' to George Dawkins 28.
Walked down to Mr. Price's – told me he had sent a Hamper to Belize.

Aug 13. Very anxious, and very busy all day, cutting Brazil Oats in Wind till 2 p.m. – then commenced Wheats in better weather in Mayes' Close, and Big Field.

Aug 17. . . . Began to carry Wheat.
Weather continues magnificent for Harvest, but Water scarce, with scarcely any Turnips, in the Country.

Aug 20. . . . Boddington murdered Beckley in Blenheim Park, on Augt. 12th by striking him on the Head with a Pick Axe – the murdered man, objected to Boddington ae 63 – paying addresses to his Daughter.

Aug 21. Mr. Price ill – thro' taking cold in riding round with me Farming, the other day.
The Brothers 'Parsloes' Slated, valleyed, ridged, & Drawn, the whole of the House & Front Porch – except the Kitchen & back Porch – weather been most favourable all the time.

Aug 24. L.M. Wynne, Rhys Wingfield – Holford Risley, &c: drove over to Shipton, and Mrs. Wynne laid the Memorial Stone of the Beaconsfield Hall – presented her with a Silver Trowel – Dr. Haigh, Revd. Barter, York, Mr. Reade, Brookes, Savidge, Lardner, Green, and several others, witnessed the Ceremony. . . .

Aug 27. Rainy morning – stopped all Harvest work.

Cricket:	Gloucestershire	v.	Middlesex
	348	110	1st innings
		233	2nd do.

Mr. W.G. Grace played magnificently *thro'* the Game, and carried his bat for 221, and afterwards bowled 10 of his opponents' wickets out of 15!!!

Aug 31. This month has been very Dry, & Scorching – should not know where to find 10 acres of *average* Turnip crop, around here!!! 17 Wheat Ricks up: & nearly thatched.

Sept 10. . . . Finished Wheat, & Bean carrying.

At night the heaviest rainfall of the year, quite [a] *Calamity* for the Harvest, in the later Districts: the Barleys will be spoiled for Malt!!! which would have been by far the most *profitable* crop of the year.

Mr. Flood retires from the Bank (London & County) 1st Jany. next.

Sept 12. Very heavy rain with high wind last night – wind contd. all day, & rain again to-night – most disastrous to the remaining Harvest, especially the Barley.

Sept 13. Rode over to Mr. Wm. Hiorns, Shorthampton: found him going on satisfactory – after his Bull had broken 3 of his ribs – thigh in 2 places – laid side of his head open, & gored the other Thigh, & sundry other Bruises in various parts of his body – will be a Marvel – if he recovers.

Sept 15. Drove Jane & Fredc. to Blenheim Horticultural Show, showed Fred. round the Gardens, Lakes, Wild-ducks, &c.

Duke Marlbro', only Son & 3 Daughters: were all with us, the eldest Daughter gave away Prizes.

Blenheim Palace, is one of the most noble Country Seats, in England.

Sept 16. Grace commences her last School Term with Mrs. Dix, Merton College, West Croydon: this morning.

W.G. Grace, the great Gloucestershire Cricketer, has this Season played 51 Innings, made 1717 Runs, & at Clifton *commenced* the Innings, & saw all his team out, carried his own Bat out for 221 runs – carried his Bat out 3 different times!!! during the Season!!!

Sept 20. The wretched weather, is spoiling much of the latter harvest, all over the Country!!! and injuring the *Condition* of *all* of it.

Sept 21. There are now, in the 3 Divisions of Oxon, 24,455 Voters as against 7649 under the late system[7]. . . .

Attended a meeting in evening, at Shipton Red Horse: and arranged the Completion of the Beaconsfield Hall.

Sept 22. Magnificent day, carrying with 9 Wagns. & 1 Cart. . . .

A letter from Lizzie – perhaps Grandpa, in Novr.

Sept 24. Carried my 10th Fairspear, 'Harvest Home'.

Began to drill 'Club Red' Wheat, in Boynal Gardens Fallows.

Hired Jas. Pratley, & Chas. Pratley, Jas. Sifford, Thos. Hanks, & Steed.

Oct 10. Had an hour with Mr. Brassey & Hounds, in Bruern woods, plenty cubs *all over* the Hunt!!!!

Nov 13. I am 56 years old today!!!

Met Hounds at Cokethorpe – Had Apples & Whiskey at Mr. Cook's.

Very damp, foggy, unpleasant day – ran a Fox to ground, and lost a Second – came away at 1 p.m.

The 'Old Berkshire' have killed 57 Foxes, in 33 mornings Cubbing!!! good start.

Nov 16. Frosty morn, only just able to Fallow –
The Rt. Honble. James Lowther – wrote me from Louth, hoping I should vote for him, in the Town Hall, on Nov. 27th.[8]

Mr. Thos. Brookes, Shipton: called & arranged a meeting for Friday next – to receive voting-day, instructions.

Nov 20. . . . Walked down & tea'd with Mr. Reade, Shipton: attended a meeting in Beaconsfield Hall – to arrange about Conveyance of Voters, &c.

Called on Mr. Price.

Nov 27. By 8.13 train & voted at Huntingdon 1 p.m.for Honble. O. Montague and reached Louth at 5 p.m. – and voted for the Honble. James Lowther – Lowther Barker drove me to Claythorpe. . . .

Dec 6. Lame of a small toe – skin chafed off – by Boot.

Mr. Victor Spencer Churchill, 3 months 'Leave' from the Army – Balls – Hunting – & Shooting.

Dec 7. Voting result in Oxon –

North	–	Sir Bernard Samuelson	4436
		L.M. Wynne	2944
			1492
Mid	–	Maclean	4327
		Visct. Valentia	4138
			189
South	–	Harcourt	3778
		Maude	3258
			520 . . .

Dec 10. . . . Mr. Hiorns going on favourably, after his *Terrible Bull accident*, of 13 weeks ago – at Shorthampton.

Dec 11. Mr. Wilkins, Chipg. Norton, made a very inflammatory speech after the Declaration of Poll – against the Primrose League – altho' his client, Samuelson, had gained his seat.[9]

Dec 21. James Sifford, rode to the 'Hawk & Partridge', Bloxham, nr. Banbury: and lead here, the Horse I bought of Mr. J.E. Parsons, Charwelton: nr. Daventry.

The new Parliament of 670 Members:

Liberals	333
Conservatives	249
Nationalists	86
Independents	2
	670

The Spectator says the Polling result, is practically a tie, between the Liberals, and the Tories, plus the Parnellites.

Mrs. Reade gave me some Pampas Grass roots: &c.

Dec 24. Drove Fredc. Geo. to Witney: Mr. Flood introduced me to Mr. Davenport, the Manager of the Birmingham Banking Co. – with whom I bank from this date – the London & County – having annulled their Witney Branch.

Christs. Eve – my 3 Sons, & 2 youngest Daughters, at Home – Alice in Belize – Lizzie in Bombay!!!!

L.M. Wynne, Nether Worton Park, sent me br. Pheasants.

Dec 31. . . . The past year, has been the most *unprofitable* for Farming, I have experienced since I began in 1852!!!

1886

Jan 1. Met Hounds at Barrington Park. . . .

Mrs. Sartoris & Dawkins & Price, joined me at the 'Merry Mouth', we inspected the Shipton 'Beaconsfield Hall'.

Jan 6. Snow windled, & fell; from 6 to 9 a.m. – mild air – no Frost: so expect it will quickly disappear!!! stopped nearly all kinds of work – for the present: 7 inches on the level.

Jan 7. Did not venture thro' the snow to Witney – Rode round by Ascott & Kingstanding in the afternoon & found the roads more passable than I had expected – but still a great fall of snow. . . .

Jan 9. Postponed Rent, till the 15 March –

Six or Seven inches of Snow over everything, besides drifts, & very Frosty & *Severe*, with N.W. wind. . . .

Jan 11. Shipton Auction, 640 sheep, tegs 7½d. lb.

Mr. Price resolved to prosecute Smith, for Partridge shooting, against his Rick.

Jan 12. The new M.P.'s meet to-day for swearing-in – and Her Majesty opens Parliament in Person, on the 21st., to the great delight of her subjects.

The ground has already been covered with snow, 7 days.

The Revd. Herbert Marlow Ward, wrote to say that Alice was dangerously ill.

Jan 19. John Chas., Jane & Isabel, attended a Concert, in Beaconsfield Hall.

Wiggins, & Steptoe, &c: again attacked 'England', the Policeman, at Leafield: & were imprisoned 3 & 7 weeks – 2 others fined for rioting, and Smith, Shipton, fined &c 35s. for Partridge shooting, & a double certificate to follow!!!

Jan 21. Snow fell – & suspended Thrashing, and Ploughing, &c:

Alice, confined of a Son, at St. Mary's Rectory, Belize, British Honduras on Dec. 22nd. 1885 – and very glad we were, to hear, she is going on favourably.

Queen Victoria opened her 11th Parliament in Person, this day; to the great delight & satisfaction, of *all*, her subjects.

Jan 24. . . . Met 'King' the Policeman . . . who expects a heavy day at Chipg. Norton on Wednesday, 6 to convict, for being riotous, & attacking the Policeman 'England' at Leafield, & another for Partridge shooting at Shipton – Smith.

Jan 27. . . . Can scarcely Farm – Hunt – or shoot: never had so little Labour, no Turnips, & altogether, the most idle Autumn, & Winter, I ever experienced.

Jan 28. Spent 2 hours at Witney – dull, depressed market: Politics all the Talk – Marq. Salisbury resigned – Mr. Gladstone is expected [to] form the next Ministry – every prospect of an unsettled year – with another General Election before it ends!!! is the Public's Opinion!!![1]

Feb 2. Walked to Mr. Price's – Mrs. Samuda, had an operation performed in London – brot. on thro' over-exertion &c &c.

Feb 8. A large meeting of Socialists, & unemployed, met in Trafalgar Sq: hence to Hyde Park – turned riotous – passed the Clubs & in 1½ hour, did £50,000 Damage.[2]

Feb 10. At 2.40 p.m. two of four Wheat ricks fired in Big Field: 200 yards off the Brick-kiln, & 22 yards from the Highway; – Believe it to be some mischievous wretch, passing by, and thought he would like a Blaze – foggy at the time – we hung a large sheet up the side of the 3rd rick, & kept it saturated with water, & *no wind*, saved the other 2 ricks – they stood 8 yards apart: only – insured in the Norwich Union.

Feb 11. . . . Filled up, & answered 10 questions, about Labourers Allotments, from the Crown.

Great Distress, and *Scarcity* of work, nearly *all* over England!!!

Feb 15. Called at Mr. Bliss's, Chadlington: he was suffering from a fall over ruts, from his Horse, while out with the Hounds on Saturday – called, & Mr. Neild, Deane: accd. me to Heythrop, Mr. Brassey gave us Orange Brandy – found 1st Fox in New Plantations – & ran to ground, in Mr. Neild's offyard – ran & lost 2nd & 3rd Fox – nice sporting day – Whiskey with Mr. Bliss as we returned – Mr. Price & I.

Feb 18. My loss by Fire – valued £175 and £1.15.6d. expences – 90 qr. Wheat 30s. – 2s. per qr. deductions for Thrashing, Dely. &c – 22 Tons Straw at 35s. – 22 acres at 5s. for chaff & cavings – Thatch – altogether, am satisfied.

Feb 25. Lord Randolph Churchill, making telling speeches at Belfast, &c.[3]

Wrote, and 'Papered' Alice & Lizzie.

John Chas. & Robt. Wm. attended a Ball, at Junction Hotel which passed off most pleasantly – 75 attended.

Feb 27. Hauled Wheat to Machine 5 days consecutively, & finished carrying Harvest!! Snow on the ground, which drifted on the 6th January ult!!!

Terrible Depression in Farming & Trade – good quality dry Red Wheats 27s. to 29s. per qr. – Blk. Oats, 40 lbs. bushel 18s. & 19s. on rail.

Prince Albert Victor, hunting with the Bicester Hounds: and visiting Lord Chesham, at Stratton Audley.

March 5. Very keen, frosty weather, with wind, & snow squalls – every appearance of a late-growing Spring – no March grass: hope to have an old-fashioned Spring & Summer!!!

Selling Red, sound Wheats 28s. on the Rail – Ruination.

March 10. Mr. John Cook's Stock & Implement Sale, at Bruern – Ewes & strong lambs 38s. to 43s. in-lamb Ewes 32s. to 36s.!!! Best ewe tegs 38s. to 42s. the Best in-calver £19 – yearlings £4 to £4.15s. Strong aged Cart Horses £10 to £14 – *Ruination* quitting Farms. . . .

March 18. Tea'd with Mr. Mrs. Worley, Witney – saw the Salvationists!!!

A complete Thaw, at last.

March 27. The Duke of Marlboro', is about to offer 408 Paintings for Sale, in London.

Mr. Newman, Clanfield: failed: bot. land, and it has ruined him – & *hundreds* of others: very good farmer. Wind & rain – mild.

March 30. Mr. Ruston's men, on strike, at Lincoln:[4] what Idiots they are, and what a curse 'Strikes', have become, to manufacturers!!!

March 31. A hindering month – no Spring Corn drilled: never seen the Spring seed time, so backward, since I came into Oxon.

April 6. Tea'd at Mr. Abraham's – with Price, Timms & Tarrant called.

Recd. a letter from Mr. Clutton reducing Rent to £985 from Michs. 1884 to Michs. 1888 – besides a Refresher of £200[5] – £455 to *deduct* from £577 – the ½ years Rent due to-day!!!

April 8. Witnessed Morton Steeplechases – Lunched with Mr. Henry, and Albert, Brasseys.

Godman won the Hunt Chase, on Silvertail.

Dined Mr. Price's with Mr. John Parsons.

April 17. Hounds met to Finish the Season, at Ascott, found in Hensgrove – ran to Widford, Swinbrook Valley, past Kite's Bridge, nearly to Minster Mill, up into Standridge, & killed the 73rd Fox – found 2 or 3 in Widley, ran in & out, & around the Cover, in cold wind & drizzle till 4 p.m. – left & tried Shipton Barrow blank – so ended Mr. Brassey's 13th, and *Best* Season of his Mastership.

Mr. Henry Brassey introduced me to his eldest Son, who retd. from Eton yesterday – the Honble. Miss Verney, was out.

Sir Wm. Harcourt's Budget, 8d. Inc. Tax contd. and £16,000 remitted to small Brewers, & 'as you were'.

April 19. Primrose Day – met the Cotswold Hounds at Puesdown Inn, 2½ miles beyond North Leach – rode from Fairspear to North Leach in 2 hours – via Taynton, & New Barn. . . .

April 21. Mr. Gladstone's Irish Home Rule Bill – and Land Purchase Bills likely to be a failure.

Mr. Smithson, ae 45, Witney: died – the Popular Station Master.

Both the Gilletts of Tangley, & Crows Hall, Bankrupt.

April 28. Why is Mr. W.E. Gladstone, like a Potato? because he is always found in an Irish Stew!!!

April 30. Walked by the Water-wheel to Shipton Auction – Mr. Hiorns's last 28 tegs averaged 39s. in Wool only!!! Gelt ewes 33s.

Called with Mr. Bliss at Mr. Leake's, & we tea'd with Mr. Price.

A cold ingenial April – Snow in Wales, & Scotland, now. Mr. Fox had to return from Salmon Fishing to Sheffield till the Weather proved milder.

Mr. Wingfield, bot. Tangley, for £12 per acre: and 'Crow's Castle Farm', is offered him, at £13 per acre, all *adjoining*, Land never been cheaper, the last 100 years. . . .

May 4. Rode over to Woodstock – & met Mr. Collis, accd. him over the Orchid Houses in Blenheim Gardens – Messrs. Bethel, Gask, Martyn, Gillett, Thompson, &c joined us at Dinner, 6½ p.m. – King's Arms.

The Orchid, is a *rare valuable* Flower, for sale in Covent Garden, a large sum will be realised out of the Blenheim varieties.

May 5. Earl Redesdale died on 2nd ult ae 81, buried at Batsford: Bertram Mitford, succeeds.

May 6. . . . Telegram to summon Jane to Southport – her Mother being dangerously ill.

May 10. . . . Mrs. Fowler's corpse travelled from Southport, by rail to Brigg, & by hearse to Elsham, where she was buried by the side of her husband, who died in Augt. 1846!!![6]

May 14. Great storms of wind, rain, snow, & hail in various parts of the Country – the Lake Districts, & North Wales covered with snow. 3 days rain at Birmingham & Chester.

May 18. Grace & Isabel, accd. me to a Concert, at Beaconsfield Hall, the surplus funds, to be applied to repairs of the Church Organ.

Coningsby Disraeli, performed & sang, to his Banjo – Mrs. Samuda, Mr. Mrs. Misses Barters, Mr. Kyffin, Mr. & Miss Hall, and the Shipton Choral Society, also assisted, altogether a very nice evening's entertainment: for 1s. each.

May 19. Showers, & Sunshine – Very growing!!! . . .

Mr. Rhys Wingfield, has. bot. Tangley Farm & Covers, and 'Crow's Castle', at £12 per acre, *all round* – Cheap!!! £13,500 & £1920.

May 24. Queen Victoria 67 years old this day – has 7 children living, 30 grand-children, & 6 great grand-children – been a Widow 24 years, & the 35th occupant of the English Throne, since the days of William the Conqueror!!!

Tea'd & walked round the Court grounds at Shipton, with Mr. Reade, & attended a meeting at the Red Horse, to confirm bye-laws, for Beaconsfield Hall. . . .

May 31. . . . Been rather a wet month, 13 fine drying days only!!! . . .

June 2. Oppressively warm, & wind in South at 6 a.m. – at 6 p.m. wind in North East & rain.

Mr. Edmonds, Leicester, lectured on Irish Home Rule, in Leafield School Room.

June 3. On Monday 'Victor' a Pointer Puppy, left Moortown Stan. 8.30 a.m. and reached Ascott 4.40 sent by Mr. Barker, Caborne: Carge. 4s. through: via Lincoln, Derby, & Worcester.

John Chas. drove me to Witney – Tea'd with Mr. Davenport, the Banker.

June 4. Rode over to Charlbury & met Mr. Perrystone & Revd. Carpenter, as Comrs., and was Discharged my Sched. B. Income Tax, on £1200.7.0d.!!!

Walked down to Mr. Price's in the evening – Mr. Andrew Blake, leaves Mr. Brassey as under Steward at Heythrop, to manage Her Majesty's Osborne Estate.

June 7. . . . For the 2nd reading of the Home Rule Bill, 311; Against it, 341, or 30 Majority against Mr. W.E. Gladstone & his Party!!![7]

June 11. Ascot Race Week, 4 days of the most Aristocratic racing to be witnessed in the *World*: all the *Best* Blood of England, meet there; Men, Women, & Horses!!!

June 15. Recd. a letter from Alice – returns via 'Salerno', Capt. Shouman, via Nassau, Bermudas, & Azores; will stay in England 3 or 4 months, and evade the Hot Season, at Belize.

June 17. W.E. Gladstone gone Electioneering, into Mid Lothian – a most eloquent Politician, & Statesman – but will never carry Home Rule in its present form in Ireland.

June 24. Mr. Price accd. Jane & I, to Witney: attended Duke Marlborough's Land sale – Chase Woods Farm 411 acres, House & Premises cost £3500 – not even a shilling was bid for the lot – the first 12 lots passed – never saw such tokens of Agricultural Depression!!! . . .

June 27. . . . George Dawkins married Mrs. Sartoris, Wilcote, on 15th ult.!!!

June 28. 'Coronation Day', and John Chas's Birthday.
Mr. Robt. Abraham sent me a Cucumber, & splendid assortment of Roses from Ringwood – Report says, his son Wm., is to marry a Miss Toller, from nr. Cambridge.

July 2. Capt. Waller gave £1770 for a House & land, at Charlbury, belonging the Duke Marlborough – because he thought Mr. Chilson wanted the land for a Cemetary – *nearly* worth, about £11 [per acre] – or £1200. . . .

July 4. Electioneering all the rage, the next 10 days; the 11th Parliament [of] Her Majesty's 50 year's reign only lasted 5 to 6 months, what changes we live to see!! Quite expect Mr. Gladstone, has over-killed, the Irish Pig!!!

July 7. Mr. Wm. Wells came & bot. 14 sucking Calves, 8 Tups, & 5 ewes, & took away 2 fat Heifers, previously bot. – the scarcity of 'Keep', caused by the *Hot* Weather, after the May *rains* had been dried up by *cold* N.W. *Winds*, makes us anxious to clear out everything that is saleable.

July 12. Oxford North Division Election – 13 Polling Places, viz. Banbury, Bloxam, Cropready, Wrockston, Deddington, Barton, Enstone, Swallcliffe, Hook Norton, Chipg. Norton, Kingham, Shipton, & Charlbury –

			last Nov	
Samuelson	3677	} 493	4436	} 1492
Wynne	3184		2944	

Polled at Charlbury.
Dined with Mr. Price.
Three hours rain, & so much required, & desired.

July 20. . . . Lizzie's Son Gerard, is dead, from Teething, 8½ months old.[8]

July 31. . . . Had 3 weeks, showery, beautiful weather, for Farming!!!

Aug 3. Jane & I, by 8.18 train to Paddington, & Gunnersbury, for Mrs. Barker 54 Oxford Road, Chiswick.

To Indian & Colonial Exhibition, in aftn., very pleased with all I saw, *most interesting & instructive*, for *all classes*, to see the *great* variety of *Products, & manufactured* articles – from our Greater Britain abroad.

Aug 4. Spent the day at Indian & Colonial Exhibition, & Aquarium and Mrs. Barker joined us, & we witnessed Mr. Dicey's 'Adonis' at the Gaiety in evening – retd. from Charing Cross to Gunnersbury at 11.41 p.m.

Aug 5. Inspected the Debris after a Fire, in Whitefriar St. Ludgate St.

Saw Mr. Gladstone & Sq. Chaplin, go to Parliament to elect a Speaker and Sq. Heneage, in Picadilly.

Saw the 'Mikado' at the Savoy.

Luncheon at 'Clarence' Scotland Yard: Thos. Whaley.

Aug 6. Mrs. Barker accd. Jane, & myself, to Kew.

Left Gunnersbury, for Bishop's Road, & left Paddington 2.15, Ascott 5.5, after a very pleasant visit with Mrs. Barker, at 54 Oxford Road, Chiswick.

Aug 9. . . . This 12th Parliament of Her Majesty is composed of 196 Liberals, 73 Dissentient Liberals, 316 Conservatives, & 85 Parnellites.

Tea'd with Mr. Abraham, & Huckvale, at Mr. Price's.

Delightful rain, at night, after 4 weeks, temperate weather.

Aug 12. . . . Lord Churchill, Lady Churchill, & their only child Victor Spencer, are at Cornbury: his Lordship is over Eighty and says he shall stay, & end his days at the old Family Seat – Cornbury House.

Aug 24. Alice reached London with her Son, on Saturday last 5 p.m. – from Belize in the 'Salerno' Capt. Shouman sailed into West India Docks.

Aug 25. Alice and her Son John (4th) arrived here (Fairspear) at 5.30 p.m. – had a quick passage from Belize, started 21st July called at Nassau, & at St.

Michael's Port of the Azores, & into West India Docks on Saturday 21st Augt. 5 p.m. – after being delayed 8 hours off Gravesend in a Fog – altogether a Quick, Calm, Enjoyable journey Home. S.S. 'Salerno' Captn. Shouman.

Aug 31. Weather dry, & sunny, ripening the late Harvest rapidly – Two Reapers cut 6 days – All hands tied 5 days, & carried 1 day, with 4 Wagns. Wheat for Seed, in August!!!

Had 7 weeks most suitable Weather for everything in the Country.

Sept 1. Plums cheap, had 70 lbs. Egg Plums from Mr. Alfred Haines, Evesham, 5s. Hamper included: paid E. Greenway, Knotty Elms, Pershore 4s.6d. for 35 lbs. – Hamper included.

Sept 15. Drove to Chipg. Norton – called on Miss Watts – and the Asst. Comtee. reduced my Ascott Rateable Value, from £174.12.0d., to £126!!!

Splendid Harvest weather, Carrying with 10 Wagns. Tieing, & Thatching.

Sept 18. A *grand* week for the Harvest – Cold – Sunny – Wind – & no Fog, and only very little dew!!! the *Perfection* of *Harvest* Weather!!! . . .

Sept 22. Harvest Home – Began 23rd Augt. – Reapered with 2 Machines 17 days, Tied all up, in 18½ days – gave 4s.6d. down to 3s. per acre, for tieing & raking – 3s. per day for carrying, no Ale.

Sept 30. . . . Sept. has been a very fine growing genial month, in this locality, but fearfully stormy, & Earthquaky, in America, Greece, &c, &c. . . .

Oct 2. Well forward with Autumn work – Thatched – Potatoes up – Thrashed 8½ days, carried in 12 days – and Drilled 95a. Wheat in Hall's Hill, & 4a. Tares, in Rickyard field – Man & 4 small Boys Charlocked the Turnips.

Oct 25. Continuance of boisterous weather, with rain squalls, from the N.E.

John Chas., accd. the Leafield Temperance party, to Oxford, by road.

Oct 31. October has been Foggy, showery, & rainy – Grass, Clovers, & Turnips, have grown rapidly.

Nov 1. Met Hounds at Heythrop, had not been at 'Opening Meet' – since the 1st Novr. 1880 – when Bevercomb, refused Stable, to my Horse.

After 2 hours – ran for Great Tew – I retd. by Cuckoo's Holt via Radford to Taston, Spilsbury, Chilson, Bridge &c.

Nov 4. . . . Been showery, rainy, & Foggy, *frequently*, since Wednesday the *6th Oct.* – too much so, for lifting Potatoes – Corn Drilling – Thrashing – and Gardening.

Nov 8. Fred. Archer ae. 29 the Great Jockey, shot himself – while suffering from Typhoid fever, at Falmouth House, Newmarket – to the great regret of Thousands of racing men.

Nov 13. The Indian & Colonial Exhibition, was closed on 10th ult. after being visited since May, by upwards of 5½ millions of People.

Met Hounds at Eynsham Hall. Champagned with Mr. Mason, & his Son.

Mr. Mrs. Gask tea'd. Mr. Price in the Evening.

Grace returned after spending a Month, at Southsea, & Ryde.

Nov 24. Lord Churchill, Cornbury Park, died ae 84 – Victor becomes 3rd Baron, and inherits everything, being only Child!!!

Met Hounds Gawcombe.

Nov 30. Lord Churchill buried at Finstock, by the side of his Father, who was the youngest Son, of the 3rd Duke of Marlborough.

The 1st Funeral witnessed by Fred, & Isabel.

Dec 4. Her Majesty the Queen, came from Windsor Castle, to Cornbury House, & paid a visit of Condolence to Lady Churchill, one of her favourite Ladies in Waiting.

1887

Jan 1. Begin my 5th Diary –

Albert Victor Spencer, Lord Churchill 3rd, of Cornbury Park, married Verena Lowther, sister to Earl Lonsdale, & spent 8 first days of his Honey-moon at Cornbury.

This, Her Majesty's year of Jubilee will be memorable in History, for the erection of Charitable Buildings, Bridges, Museums, Libraries, Statues, &c great expenditure in Hospitality &c.

Jan 11. Held our annual Conservative Dinner, in our new Beaconsfield Hall, L.M. Wynne in Chair, supported by Capt. Crawford, A. Brassey Esqre., Revd. Yorke, & Barter, & 150 others – very pleasant meeting – I proposed Mr. Brassey's Health &c.

Jan 20. Drove Alice to Witney, dare not drive thro' the Windrush, at Minster, so went by Crawley: the valley full of Water – it even ran thro' Mr. Westell's House.

Jan 24. Rent Day. The S.S. 'Salerno', has arrd. in West India Docks, from Belize: Alice expects to return in it with her Son.

Met Hounds at Pomfret – killed brace Foxes – nice day's sport – much ice, and snow.

Feb 2. Received a letter from Capt. John Shouman, S.S. 'Salerno'.

Mr. Bliss, Waterman's Lodge, failed – paid 3s.6d. in £.

Feb 3. Alice & her Son, left Ascott by 1.15 train, direct for the 'Salerno' in the West India Docks – they & nurse 'Margaret', were all in excellent health. I rode to Witney.

Feb 4. At 9 a.m.: the 'Salerno' steamed out of W.I. Docks, for Dartmouth – Tenneriffe, one of the Canary Isles – Nassau, one of the Bahama Isles, & they would call also at 1 or 2 Ports at Cuba, before they landed at Belize – a long journey across the world.

Feb 19. Met Hounds Blenheim – found Fox in Bladon Wood – to ground in Palace Gardens – found in Coombe Gorse, & killed in Kingswood – found 3rd in Coombe Gorse!! ran thro' Coombe, Handboro' Quarry Cover, over the rail to Gorse, & Cornbury Forest. Home at 4 p.m.!!!

Feb 28. Met Hounds at Chapel House – found & ran to Whichford Wood – found again in Salford spinnies, & ran by Rollright Stones & lost at Compton: found at Adlestrop – ran & lost – bad scent.

March 1. Luke Shayler, accidentally shot Mark Franklin while rabbiting nr. Ranger's Lodge – the charge squandered his Brains.
Evening with Mr. Price.

March 5. Benj. Hobbis refuses to collect the Ladies Jubilee pence in Leafield on behalf of Lady Churchill!!!

March 7. Meetings held daily all over Her Majesty's Dominions, to decide upon Buildings, Festivities, & a host of other matters, in remembrance of her 50 years reign come the 20th of next June.

March 12. Had 3 weeks of Fair Weather – the land most friable and dry – Thousands of acres Spring Corn drilled – Red Wheat 32s. – Barley 21s. to 26s. – Oats 16s. to 17s.
Met Hounds Wilcote – called Mr. Sutton's – found a newly built House!!! & brace Foxes, in Ramsden Gorse – to the great delight of owner, Capt. Wynter: found next in Singe Wood – ran to Bradwell Gorse, & into Cornbury Forest, away over Witley Hill along 5 Ash valley, past Chalk-pit spinney – thro' Shrubbery & Priest Grove by Wheelhouse – into Shipton Field – & got into Drain nr. Jugg's Grave.

March 14. At Shipton Auction – Tea'd with Mr. Price – Mr. Bliss, Jas. Shillingford, & Robt. Wm. joined us – Mr. Price objected to Shillingford pressing his Friends for 'Orders', in his room!!!

March 15. Had 3 weeks splendidly dry weather, for the land, agricultural work very forward – at 10 a.m. – 3 inches of snow, but *still & mild*, from the North East.
A very amusing letter, from Lizzie, Bombay: she met the Duke & Duchess Connaught, Lord & Lady Brassey & 3 Daughters at Lord & Lady [left blank] Jubilee Ball – & also at the Races.

March 23. Met Hounds at Junction – after running brace Foxes to ground at Bruern, found in Oddington Ashes, ran an hour over Dalesford, Adlestrop, Cornwell, & killed ½ mile beyond kennels in Chipg. Norton.
Mr. A. Brassey sails his Yacht the 'Czarina', in the Jubilee Race round England, Scotland, & Ireland – start at the Nore – about 1600 miles.

April 7. The Crown posted me a £200 gratuity – reduction of rent – for year ending 11th Oct 1886.

April 9. Mr. Price and I, rode thro' Forest – called at Mr. Gask's – and met Hounds at North Leigh – found in a small cover, beyond Townsend's cover, & ran into Singe Wood – away over Chasewoods – to Blindwell Gorse, Blindwell Wood, & on to Leafield Church, turned back, and killed in Revd. T.W. Lees, Meadowfield – 1.30 p.m.

Horse lame – came Home!!!

Lord Valentia – Lady Alice Bertie & Sister, out – &c. . . .

April 13. Tea'd Mr. Reade's – and attended Mrs. Jenkins' annual Primrose League meeting at Beaconsfield Hall – Mr. Haigh ruling Councillor. Mr. Joseph Reade Junr. has Measles.

April 22. Josh. Reade, & his Sister, Mrs. Jenkins, both ill of Measles. Spring backward – not heard or seen – the Cuckoo – Turtle Dove, Swallow, or Nightingale!!!

April 23. Bad cold thro' work – heated myself – and cooled too rapidly – throat, tongue, in fact, sore & chilly all over – got the 'Creeps'. . . .

April 26. Finished thrashing last year's Harvest – out of Kingswood. A fine, heavy, still rain, from W.S.W. set in at noon – after two terrific!! *white, Rime* frosts.

May 1. A white rime every morning – with cold wind, & abundance sunshine. At Church in a.m. Sherried with Mr. Price, over *Priest Grove* Nuts, in aftn.

May 11. Grace, Isabel, & Fredc. accd. me to Yeomanry Sports in Over-Norton Park – Tent pegging – cleaving the Turk's Head, splitting the Lemon – Sword v. Sword – &c.

We looked round Chipping Norton Church, & yards, saw Mrs. Rawlinson's & her Daughter's grave!!!

May 12. Rode to Mr. Matthews's, Fifield: and accd, him to Stowe Fair – moderate sample Horses, and very *slow Trade*!!! reached Fifield 4 p.m. – walked down to the Woods by Bruern, & back on the Fawscott side. . . .

May 17. Drove Junction, and trained to Adderbury, & walked on to the Agl. Show ground, Banbury: average show – saw Mr. Edmonds, Mayor. Banbury is one of the finest, largest Country markets in England!!!

May 28. Col. Harcourt, met 3 to 400 Farmers in the Sheldonian Theatre, Oxford, who clamour for Protective duties on Corn – the case is a very serious one, for *all concerned* in *Land*!!![1]

Bot. knives and forks for B. Hobbis, on his wedding. . . .

May 30. Mr. Eagle called – on his way back from South Lawn.

90 Fire Engines met at Oxford to-day – as a Whit-Monday exhibition.

May 31. This has been a cold sunless month: with very little rain: indeed there has been very little rain, the past three months.

The Opera Comique burnt down in Paris, [left blank] persons found dead in the Debris – 3 sisters from Gloucester 'Knowlses'. Several identified by their Jewellery – only the jaw, of one person found!!! . . .

June 3. The first really *heavy* rainfall, since January 19th ult. – every prospect of plentiful crops of Grass, Clover, & Straw, & Fruit, thro' a cold spring.

Leafield Feast – dull & rainy – no waste of time!!! Mr. Price in Town.

June 10. Magnificent growing weather, all the past week –

Mitford offering his Swinbrook Estate for Sale – and the Mortgages claiming the Cornbury Estate!!!

June 13. Shipton Auction – Tea'd Mr. Price's – Mr. Franklin called.

Col. Kingscote wrote, and offered me 2 seats in a window, nr. No. 6 Whitehall – very pleased am I – from which to see the Queen return from Westminster Abbey, on 21st inst.

My 33rd Wedding anniversary.

June 17. Went via Bridewell to Witney – attended sale of land belonging to the late Mr. C. Sartoris – 19 lots sold out of 21 – Holly Court Farm withdrawn at £34 – 242 acres. Field Farm realised about £27 per acre & Church Farm £29.10s. per acre which were considered good prices!!! . . .

June 18. Decided the 'Jubilee Day', should be an entire Holiday at Fairspear – Church 10.30 Dinner 12.30 to 270 men – Cricket – 260 children tea'd – Sports – Bonfire in Leafield Piece to finish 10 p.m.

June 20. By 8.17 train to Paddington at 10.25 on to Woods Hotel, Furnivals Inn, then called & had Champagne & Lamb, with Mr. Mrs. Whaley at 63 Lambs Conduit St. – then to witness the performance of the Far West, given by the Honble. Cody & Buffalo Bill, [&] the Sioux Indians, nr. Earls Court Station[2] – Alhambra at 8 – Gorgeous scenery, & Dresses, &c.

June 21. Up at 6: walked Streets and admired Jubilee Decorations till 7 – reached No. 6 Whitehall at 8.30 – At 10 Mr. Whaley called me out, and saw Dr. Tom Robinson at his Son's the 'Clarence' Scotland Yard – we walked into Northumberland Avenue & saw the magnificent Jubilee Pageant on its way to Westminster Abbey – then returned to our seats in Whitehall – whence I had a near distinct view, of *all* in the *Procession* after they had

Thanksgiving – the Grandest sight I shall ever behold in my Life – 4 Kings & 2 Queens, as visitors, were in the Procession – Her Majesty looked hale, serene, & happy – I left Paddington 6.30 reached Home 9.30 whence I saw 27 Bonfires, including the one in my Field.

June 22. The Queen drove into Hyde Park & saw 26,000 children entertained, each child recd. a Bag containing a Meat Pie – a piece of sweet Cake, a Bun, & an Orange, and was decorated with a Jubilee Medal!!! Her Majesty retd. to Slough, & was driven hence thro' Eton to Windsor Castle, where a statue of Her Majesty was unveiled!!!

June 25. The Weather the past *3 weeks*, has been dry, & sunny in the extreme, the Foreign Visitors would be favourably impressed with our English Climate – all kinds of Vegetation, never looked better.

July 1. Extremely parching, the grass looks brown as mid-Harvest: will be very injurious on all Sand & Gravel soils.

Dr. W.G. Grace, is the best all-round Cricketer in the World, his performances this Season are superlative!!! 116 and not out, & 92 2nd innings v. Surrey. . . .

July 7. At Witney – Wool 24s.

Dr. W.G. Grace, the great Gloucestershire Cricketer, made 92 in his first innings v. Yorkshire, and carried his bat out in 2nd innings after making 186!!! he is the best all-round player in the *World*!!! and so far this season averages 76 per innings!!! unprecedented!!! . . .

July 16. . . . Dr. Hayward, Stow, died on 3rd ult. after practising more than 50 years on that Bleak, exposed part of England – he enjoyed Fox hunting. . . .

July 18. Dr. Hayward has left his niece's Son £10,000 when 23 years old.[3]

July 21. At Witney – Red Wheat 31s. – Wool 23s. – The Country daily becoming much browned up thro' the Continued Drought!!!

July 30. . . . Magnificent weather, but too Dry for Farming: scarcity of Grass, Clover Eddish, & Turnips!!! & Spring Corn straw – and Water – in many Districts!!!

Aug 1. Began Harvest with 2 Reapers, in Mayes Close Oats, & Big Field 28a.

. . . Mr. Morris commd. bringing letters from Ascott. . . .

A Balloon from Cheltenham alighted at Ast[h]all – a military Captain & a Companion packed it up in the wicker Basket they had ridden in – and retd. by rail from Witney – I thought it would have settled in our Big Field!!!

Aug 3. Rode over to Mr. Adamses Hailey Manor, found him very busy 'Fagging Wh[ea]ts', Mr. Whale was there.

Mr. L.M. Wynne bids £1500 a year for Cornbury House and its Belongings – I hope he will take it!!!

Aug 13. Another fine Dry week – Harvest progressing most rapidly *all over* England, 3 weeks earlier even in Cumberland than usual – owing to the very dry season, Windermere Lake 5 feet lower than ever known in this generation!!! *Woods, Forests, & Moors*, are being ravaged most injuriously, by Fire, in Wales, Scotland, America, &c.

Aug 19. Thrashing wheat – thrashed a few Vetches & made day up with wheat on the 13th ult. – which was probably the *earliest* thrashed Wheat, in my 35 years farming!!! . . .

Aug 23. Carried my 35th Harvest – began on 1st ult. & began as to-day, *last* year, so carried 2 Harvests in *one* year. Sold 60 qrs. at 28s.9d. Wheat – Dry and very good quality!!!

Aug 26. Bampton Fair, the weather is fearfully sunny & Dry – the land baked out, cannot plough – water scarce – springs failing.

Sept 10. Filling up the Quarry in Shakernook – no stone left. Thoroughly repairing Priest Grove Wall – next Hambidge. Labourers 9s. per week in many parishes around here!!! Free Trade!!! *Ruination, & Starvation*!!!

Sept 24. Been Drilling Vetches, & Wheat, all the week!!! very early; but goes in admirably!!! and on the heavy Fallows i.e. Farthings, Rickyard Field, & Leafield piece. Agricultural work very forward. Best Red Wheats 26s. to 28s. – Barley 22s. to 30s. – Oats 11 st. 15s. . . .

Oct 10. Shipton Auction – Store stock ruinously *Low*! was bid £25.10s. for 6 1½ year old Heifers!!![4]

Lunch with Mr. Eagle at Mr. Price's.

Oct 11. Lady Brassey died of Fever in the 'Sunbeam' on Sept. 14th ult. – contracted on the Australian Coast, & was buried at Sea!!! between Darwin & Algoa Bay!!! "Sic Gloria transit mundum".

Oct 30. Mr. F. Taylor accd. me to Church – looked over the 'China Shop' on our way, & saw the great variety of articles made from the cold mixed Oxford clay – but all burn 'red'.[5] . . .

Nov 1. Fine, still, light night: a Blood-red sunrise – a Gale at 9, with much wind, & rain, all the day. . . .

Nov 12. Met Hounds, & had Champagne, at Eynsham Hall – found in Pinsley, ran by North Leigh Common, Holly Court, Bridewell, Wilcote, Finstock, into & around Cornbury Forest till we killed 1½ hour – found again in Bridewell – ran to drain – out & actually to ground again – left him.

Nov 18. *This* day, Mr. Mrs. L.M. Wynne, leave Nether Worton, for Cornbury House, for 14 years!!! no doubt they will benefit the locality very much!!!

Henry 3rd and Geo. 3rd are the only 2 Monarchs, who ruled longer than our Queen, who I hope, will surpass *all* her predecessors!!! . . .

Nov 23. By 8.17 to Oxford Grand Conservative Gathering, presented an address for Leafield, Wychwood, Finstock, & Langley: to Marquis Salisbury who is Chancellor of the University – spent an hour at Indian Institute – saw Photo's of many of the *wonderful* Buildings in India!!!

Walked round Christ Church Dining Hall – which contains the Likenesses of eminent men who were educated there, there are also 28 Likenesses in the Vestibule. . . . .

Nov 29. Attended meetings at Leafield – Water Supply left over – & Voly. School rate, 2½d. in £.

Looked over new Stables at Bruern, belonging C. Samuda – found them very substantial – Mr. Groveses contract £2300. Laughed at *Ensilage*, & Calf *Rearing.*[6]

Nov 30. Had very various weather during the month – Hounds stopped a week by Frost – only small rainfall.

John Scoveall married Freeman-Mitford's French Governess short time ago!!!

Penson of Ascott married Miss Edwards of Ramsden.

Mr. Gregson farming Potters Hill – but he cannot teach us to make money at present prices.

Dec 2. Moreton in the Marsh, Town Hall, opened: cost upwards of £10,000 – Built by Freeman-Mitford in remembrance of Earl Redesdale.

John Chas. met Hounds at Farmington.

Dec 5. Mr. Gregson drove me all round the Farm, and I showed him its Boundaries, Buildgs, &c.

John Chas. had quite a good day's sport from Boulter's Barn – Lord Elgin, kicked on the leg by Samuda's Horse – his Lordship is now lying up at Shipton Court, till sound again.

Dec 14. . . . Met Hounds at Adlestrop – found in Oddington Ashes & lost in Icomb – found in Adlestrop Hill, ran to Little Compton, up the Hill to Cross Hands on nearly to Rollright Stones turned, & ran over the Country to Barton Grove &c. Called with Mr. Ed. Bliss, at his Brother Jameses, at Rollright, as we returned.

Lord Thos. Brassey, & Family, arrived in the 'Sunbeam' to-day at Portsmouth: all well: after over a year's journeyings in India, Australia, &c – during which time his Lady died of Fever.

Dec 16. Mild weather, with frequent showers, & rains.

Fredc. George goes to Thos. W.W. Hopwood's, Louth, Linc: on Jany. 23rd to attend Louth Grammar School!!!

Dec 18. Walked all round the farm – Aldin's Foreman, drowned himself. Fine open weather – for sheep – & ploughing.

Dec 26. Miss Gilbert, Lincoln, sent us the most elegant Xmas Cake we ever had in the House – it was flavoured with Otto of Roses, and ornamented in Green, White, & Gold.

Dec 29. Walked to Mr. T. Brookeses, Shipton: & John Baufran's, Milton, tea'd with Mr. Price on my return – Mr. Reade joined us for 2 hours – arranging allotments for the Poor.[7]

Dec 30. Too hard to plough since Monday – no snow, no rain, dull thaws during day, equal to the freezing at night – no wind: thick heavy leaden atmosphere.

Dec 31. Much of the past year, been the *worst* on *record* for *Farmers* – the weather set in, sunny & hot from 10th June for *3 months*, consequently no after-maths – half a crop of Spring Corn, – Grass burnt up – a fifth of a Turnip Crop, and the young Clovers, more than half scorched up – *three-fourths* of the Wheats, good crops, at 28s. to 31s. – for Reds – Barley 23s. – Oats 15s. – Ruination to *Landlords, & Tenants*!!!

1888

Jan 1. Fetched an ivy leaf into, before I took myself out of, the House – so expect a better year's luck!!!

Frost, with cold cutting wind – fine seasonable day on the whole.

Jan 2 & 3. Attended Shipton Labourers' Allotment meeting – Farmers, & men – *all* attended: Elm Tree Walk, chosen: 36s. per acre.

Supped with Mr. Price.

Jan 11. Met Savernake Forest Stag Hounds and Stag at Lechlade – most annoying – a Fog – and did not clear off.

Capt. Riley, gave many of us, Lunch –

Admired their Yellow Coats, Vests, Blk. Caps, *small* white cord breeches, white ties – Brown top Boots – the Deer Taker was in Blk. Velvet.

Jan 14. Had much Fog all over England during this week – could not see the Forest during the last 4 days: no rain, or snow, scarcely!!!

Jan 23. Fredc. accd. me by rail to Worcester, heard chanting in Cathedral, went on to Birmingham, Derby, Nottingham, and admired Lincoln Cathedral, reached Louth at 5.10. Dined with Revd. W.W. Hopwood at the Lodge – stay'd night.

Jan 24. Looked round Grammar School &c – left Fredc. to be examined & commence Term. . . .

Feb 14. A windy, snowy day; a regular American 'Blizzard'; 6 to 8 inches on the level – if the wind had not drifted it – 4 to 10 feet deep, in certain places.

Feb 15. Wind, & slight snow squalls, all day: I always dislike snow as a Farmer – Sheep suffering, & more expence – with *no* Profit!!!

Feb 17. Roads been dug out, past 3 days, no road from Leafield, to Shipton, for Wheels, till to-morrow at noon!!!

Feb 25. Eleven men digging snow, from Brazil Hovel, towards Ascott. . . .

March 6. A thaw set in – ground been covered with snow – since the evening of Feby. 13th!!!

March 9. At 8.30 a.m. – Emperor Wm. 1st. of Germany died, he was born on March 22nd 1797: lived to be the most magnanimous Sovereign in Europe – his only Son, Fredc. Wm. & our Princess Royal, now become Empr. & Empss.

Tea'd with Mr. Price – Mr. Savidge came in.

March 17. Hounds met Ascott – snow on ground, Wind, very cold, bad scent: found 4 Foxes in Hensgrove, &c – could only run them about a mile!!! . . .

March 22. Jas. Dores – who has cut down my Covers and plashed hedges for 13 years – buried at Leafield this afternoon, his 7 Sons and only Daughter, attended his Funeral: a real born old Forester – *clever* at wood cutting, & hedge plashing.

Our Easter's meeting – Mr. Addy & Benj. Hobbis, attended, & with myself, hold the parish offices.

March 24. Called & saw L.M. Wynne, at Cornbury – met Hounds at Kiddington Gate – found in Glympton Gorse, after a good 55 m. to ground in North Aston – large 'Field', 10 Ladies, viz. The Honble. Mrs. A. Brassey, & Daughter: the Honble. Miss Dillon, and Mrs. White – the Honble. Mrs. A.W. Hall – Mrs. Kingscote – Miss Alice Mason – Miss Wynter – Mrs. Callendar, Mrs. Wynne –

Sent £492.10s. to Whitehall.

March 26. Three inches snow – with frost.

Mr. Goschen introduced his wonderful Budget, 1d. in £ off Income Tax – making it 6d. on Sched. A. Farmers are to be charged under Sched. D. Pleasure Horses & Ponies £1 each – 5s. per doz. on imported *bottled* wines – and *many* other *items.*

March 28. Snow again during night, & rain till 10 a.m.

I rode away from here (Fairspear) 10.18 reached Lechlade 11.45 where I met the Savernake Forest Stag Hounds, and the Famous Stag called 'Stonehenge' – uncarted him nr. Mr. Davis Mill 12.15 he ran mile towards Faringdon & turned over Kelmscott, & ran thro' Langford, Filkins, by Nashes Gorse, & to Southerop, Hatherop Park, Sir Thos. Bazley's beautiful Grounds, Water, & Castle – on to Quenington, came to a check on Mr. Hobbs' Farm, in Maiseyhampton, lost 15 min – then on for Poulton, where in wind & rain, I came away with Mr. Wakefield of Lechlade, who directed me the way to Westwell &c – the Hounds ran 7½ miles further, and the Stag was taken nr. Oaksey 3½ miles beyond Cirencester – if rain had not fallen should have seen the Finish, and stayed night at Cirencester – 13 couples small Hounds – 'Stonehenge' looked in Fine Condition, very wild, & is Dun-coloured; ½ a mile away, he looks a cross, between a Camel, and a Dun Coloured Horse.

April 5. Received a letter from Whitehall, remitting me half a year's Rent, for the 11 weeks scorching, of last summer – very liberal, but really much needed!!!

Met Hounds at Woodstock – finished Cornbury.

April 6. Attended Moreton Steeple Chases – R. Hambidge won the Farmer's Cup – John Chas. rode Mr. Davis' 'Ringcraft' 4th in, 8 ran. A. Rushout, A. Brassey, Captn. Thursby, & G. Dawkins ent[ertaine]d me.

April 13. Met Hounds Cornbury Park – ran 2 or 3 Foxes up and down the Forest all day, & finished the Season, which has been very unsatisfactory for really Hunting men – thro' Fogs Frosts & Snow in abundance.

April 17. Accd. Mr. Price to Williamstrip the seat of Sir M.H. Beach – saw Hatherop Castle – Coln St. Aldwyn's – and the Barnsley Covers, retd. back by Aldsworth, altogether had a nice ride round. . . .

April 19. Isabel was driven to Potter's Hill, for her *first* school attendance with Mr. Mrs. Gregson's Governess – Miss Evans.

Too showery even to drive to Witney.

'Primrose Day', in memory of Earl Beaconsfield.

April 24. The Queen reaches Berlin at 9 this morning, when for the first time, she will behold her eldest daughter, as Empress-Queen!!!

April 27. Her Majesty reached Windsor at 7.15 p.m. direct from Berlin, she has had a grand visit – and is said to be in the Best of Health – truly she is the most respected Monarch, in the World!!!

May 2. Fine dry still morning at 6 – high wind at 9 – afterwards, frequent, heavy showers.

Voluntary School meeting broke up, abruptly: only Revd. Lee eligible to vote – Bateman, Addy, Abraham, Hobbis, & Self – 'not in it'.[1]

May 4. A Loan Exhibition at Town Hall, Charlbury: to be opened by L.M. Wynne – I sent 80 coins – 2 Cocoa Nuts – Gulf Weed – Hawk's-bill Tortoise – 3 Orange-wood Sticks, wild Turkey's Tail – Peregrine Falcon – & case of 5 Hawks. Open 7th. to 11th. inst.

May 7. Attended a Loan Exhibition in Charlbury Town Hall – opened by L.M. Wynne Esqre., Cornbury House – the 3 Brothers Albrights, Holway, Bowley, Smith, Capt. Waller, Wilcox, Revd. West – Paine, Midgley &c &c

– a very nice show of all sorts of things, gathered from various parts of the World.

May 16. Called on Revd. T.W. Lee, and paid School Rate – £581 – £63.10s. – & 9 Cottages £21 [total] £665.10.0d. at 2½d. in the £ – is £6.18.8d.
Drilled my 2nd day's Turnips – Imp. Gr. Globes: very early.
Bunting painting exterior House, & Premises: 3s. per day.

May 23. Walked to Mr. Price's 4 p.m.
John Norton, Milton: in a rage, slew his wife, with an iron bar.

May 24. Queen Victoria, 69 years old!!! Her Grandson, Prince Henry of Prussia and her Grand-daughter, Princess Irene of Hesse, were married in Berlin to-day: in presence of the Emperor Fredc., & Prince of Wales, &c.
Rode to Witney.
A Nurse to Jane, from Oxford, yesterday.

May 31. David Bunting – the painter – died ae 32.

June 6. Nurse 'Lawrence', from the 'Acland Home', here still.
Rode round, & saw Mr. Bennett, & his Pheasants, on Ranger's Lawn, 5 to 600: he has already sent off 20 Cub Foxes, out of the Forest – *one* litter of *10* out of *one* long straight 'Earth'.

June 12. Jane seriously ill – 34 years to-morrow, since marriage –
Mr. Mace, will quit Langley Farm, at Michs.: only £50 Rent paid the past 3 years!!!

June 14. By 8.17 train, reached Ascot 11.30 called on J. Comins – saw the Hounds – Drags drive into Enclosure – Earl Coventry Mr. Harvey (the newly-appointed Huntsman) Comins &c. drive up the Course – sad news from Berlin, about Empr. Fredc. prevented Royalty attending.

June 16. At 11.15 a.m. yesterday, died the Beloved Emperor Fredc. of Germany, only surviving his aged Father, since the 9th. March *last* – a quarter of a year's Reign!!! was born Oct 18th 1831 – leaves 3 Sons & 4 daughters by our Princess Royal – reigned 99 days.

June 22. The Revd. H.M. Ward, Mrs. Ward, & Son: all safely reached London, from Belize – via Jamaica, Boston, & N. York. . . .

June 26. Mr. Southwell called, about Leafield Allotments, &c. . . .
Robt. Upton, to be hanged, for striking his wife dead, at Milton.

June 27. Nurse 'Lawrence' from 'Acland Home' – returned back after 5 weeks service here – 37 Wellington Sq: Oxford and Mrs. Meadowcroft, came from Hackness, in her stead.

Much rain in aftn. & evening.

June 29. The Emp. Wm. 2nd of Germany – begins his Reign amongst great popularity, with all classes – our Queen will be proud of her Grandson!!! . . .[2]

June 30. The past month has been very favourable for Farming around here – frequent rains – have ensured a crop of Turnips – and abundance of everything else: a great contrast to last year: 40 acres Turnips singled 3s.6d. 108 acres Drilled and only 10 acres more to Drill. Gooseberries & Potatoes, been usable a week.

July 2. Rain at 5 a.m. and much rain during the day – enormous quantity of Grass & Clovers, cut around the locality – much of it spoiling.

Mr. Shillingford bot. wool at 21s. – the 13th clip – all of which he bot. – since I've been in Oxon.

Three inches rain fell in 24 hours, at Bolton.

July 14. Began cutting 47a. Big Field piece Clover on *19th June*, rains stopped me 2 or 3 times – and to-day *began only*, and carried 7 a.m. to 7 p.m. Only 6 or 8 loads left out, rain set in again 7.15 p.m.!!! much of the Fodder of the *whole* Country, very much injured – and in a many instances will be nearly worthless!!!

July 17. J. Mitchell expects Sch. B. Inc. Tax: out of my 'Gratuity'. Rain every day – much of the Hay spoiled all over the Country.

July 19. Attended Bazaar held by Mrs. Rees-Jenkins at The Lodge, Shipton – Grace & Isabel accd. me.

Mr. Pierce's eldest Son dead – Miss Bishop died in her morning Bath, at Shipton College – Rain at 7 p.m. – Tea'd Mr. Price's.

July 20. Frequent showers – so very different to last year.

Wrote 'Crown' about Sch. B. Inc. Tax – J. Mitchell, furious!!! . . .

July 28. Mr.Gregson walked over for an hour, in the Evening.

The new Duchess Marlborough, has had 3 days at Blenheim: and is about buying Earl Brownlow's House, in Carlton-house Terrace, for her Town House – in which she intends to hold 'Great State' next Season – Nellie Warren Price – Mrs. Hammersley – and now Duchess of Marlborough: are her three names during the first 34 years of her life!!!

July 31. Shipton Fete – Mr. Price entertained my 3 Sons, & Grace & Isabel!!

Rain nearly every day during the past month.

Aug 1. Mr. Gregson & I, went by 8.18 to Oxford, & 11.15 to Thame – thro' Littlemore – Wheatley, and Tiddington, were driven thro' Thame & Long Crendon, to Mr. John Treadwell's, Winchendon: his 62 Rams averaged £17.17s.2d., they are the most superb flock of Oxford Downs in England.

Aug 2. John Chas. & Fredc. Geo: to Witney.
Mr. Josh. Druce, Eynsham: Bankrupt.

Aug 3. Mrs. Meadowcroft – left.
Mr. Cheetle, met Mr. Wilcox, and gave his opinion about my wife's affection of the Liver, Kidneys, & Heart.

Aug 4. Mrs. Hamilton, came: to nurse Jane.
A great 'Hay' carrying day, over much of England!!! after showers, & rains, for *6 weeks*.

Aug 6. Drove Fredc. Geo: to Blenheim, and lost him – saw round the Pleasure Grounds – 2000 Vol[untee]rs drilled in the Park, and attended the Foresters' Fete.

Aug 10. Finished carrying Clover – 6 Ricks in *good* Condition, and 2 in moderate 'order'.
Clover and Hay Harvest, is the most anxious Farmer's item, of his year's Labour: but it is marvellous what you can now perform, with a *Grass Mower*, and *Horse Rake*!!!

Aug 11. John Chas. drove Grace, Fredc., & Isabel – to Oxford. Robt. Wm. by 12.48 train to see Mrs. Ward at Finsbury. Mrs. Hamilton left!!![3]
The Railway thro' Charlbury & Ascott, was opened in 1852 and the *last* 'Forest Fair' was held in 1855 – when 10 Excursion trains brot. people, to Charlbury Stan.!!!

Aug 15. Drove Mr. Price to Mr. Foxes sale of Oxford-downs: told me: his wife had sent him a £500 cheque from Torquay – Mrs. Bowdry left Mrs. Price £3100, and *all* Plate & Furniture: and Mr. Price £500!!![4]

Aug 27. John Chas. drove Fredc. & myself to Bampton Fair – large show Horses – very slow low sales – called Mr. C. Hobbs' – Home at 2 – there at 10.
Attended a Primrose meeting, at Mr. Reade's, Shipton, in p.m.

Aug 31. 9 men began Pea cutting in Mayes' 12 acres: last year I began Harvest, on 1st. Augt. & *finished* on *23rd*!!!
A *whole* fine, fair day: with every appearance of a 'clear up' . . . A dull, cold, wet, stormy Augt.

Sept 1. *Began* Reapers in Fernhills Wheat – a brilliant sunny day – the *Best* day for weeks!!!

Shot at, & killed 5 Partridges – a double, and three single shots.

Sept 11. Miss Annie Wilson, Kirmington.[5] Grace & Isabel were driven with me to Blenheim Horticultural Show.

The Duke and Duchess Marlborough and his only Son – Marquis Blandford, walked round and the *new* Duchess was introduced to the Elite.

Sept 22. A whole *fine*, fair week: carried to this date, 250a. Wheat Barley & Peas: 150 acres Oats to carry, & 100 acres to cut.

Mr. Davenport, on Bicycle from Witney.

Sept 24. Rain – Miss Annie Wilson, Kirmington: left for London. Jane & Robt. Wm., travelled to Bournemouth, for a fortnight.

The Australian Cricketers, have played 40 matches in various parts of England – won 19, lost 14, and drawn 7 matches: Turner was their *great* Bowler.

Sept 25. Wind & rain, after a 2 days Fog from the N. East. Burford Hiring Fair, for Carters, and Shepherds. Ploughing, & thatching.

Oct 10. Harvest Festival at Leafield!!! the Revd. Payne of Coggs, assisted our very respected Revd. Lee.

Oct 13. Carried 'Harvest Home', in grand weather, and *earlier* than we expected, after the rains in July, & Augt. . . .

Oct 20. Hounds had a day's cubbing in Cornbury Forest for the 1st time of 20 years!!! did not kill – but I very nearly shouted a Cub into the Pack – 66 Hounds out!!!

Had some talk with Mr. Mrs. Wynne – Mr. Mrs. Dawkins – Capt. Evans, & Wynter – Mr. Brassey – &c.

Brilliant sunny hot day. . . .

Oct 23. Attended Leafield School Comtee. Meeting – Ld. Churchill, Richd. Ferriman, W. Webb, & myself, were appointed the 4 Comtee. men – according to an old Act in Leafield Parish – formed many years ago: next Voly. Rate 2d. in £.[6]

Oct 31. . . . October been a fine mild *Dry* month.

Nov 2. . . . A regular dull drizly November day, with rain at night: which is much required in many districts, as the Water supply is becoming short.

Benj. Hobbis has Oats to *cut* at Leafield, and many fields of cut Beans are in Stock all around here!!!

Nov 3. Met Hounds Eynsham Hall – had a Pheasant, & Champagne breakfast – Drizly misty day – Good sport, killed.

Nov 11. Very mild, dull, Foggy weather; with night rains – Oats out in Leafield, & Beans in Shipton!!! yet. Very much Wheat to plant in various parts of the Country.

Nov 14. . . . Mr. Matthews elected County Councillor for Fifield, Leafield, Wychwood, Milton, Langley, Lyneham, Idbury, Bruern &c.[7]

Nov 20. Rode to Kingham Church in an hour, & to Golden Cross Inn, in ¾ hour – Met Lord & Lady Willoughby and the Warwickshire Hounds, ran as if for Brailes, turned and lost, amongst Sheep, in a grass field, on the north side of Wolford Church – found again in 'Aston Hale', ran over the Rail into Batsford Park – I left, and inspected the new House being built at Batsford, reached [Home] in *high* wind at 4 p.m. – very pleased with my day!!!

Nov 23. Finished Wheat sowing in Big Field.

Lent B.Hobbis 4 Waggons, 4 hands, & 7 Horses, and carried his 13a. Oats at Leafield.

Looked over Leafield School repairs with Revd. Lee – Mr. Price walked up to Tea.

Very mild windy weather. . . .

Mr. Hambidge fast failing –

Mr. Beauchamp & Cresswell, farming at Potters Hill.[8]

Nov 27. . . . The Revd. Majendie, Woodstock; foolish enough to refuse the Ringing of Woodstock Bells, when the Duke Marlbro' & his Bride, arrived at Blenheim.

Nov 29. The newly built 'Birmingham Bank', opened at Witney to-day, which is quite an improvement to the Town.

Mrs. Ward (Alice) came from Finsbury Park, and accomd. me Home at 4 p.m. – to see her afflicted Ma.

Nov 30. L.M. Wynne, having 4 days shooting at Cornbury: yesterday & Wednesday, they shot 587 Rabbits, 350 Phts., & 3 Hares: & 3 Woodcocks.

Mrs. Ward returned to London by the 4.54 train. A mild gentle rain, all day!!! – a very *rainy, windy month*!!!

Dec 4. Mr. Franklin called for Revd. Barter's Tithe.

Jane continues very ill: from Heart, & Liver, affection.

Dec 5. Tea'd Mr. Price's.

Great opposition, likely to arise, in electing County Councillors in January!!!

Dec 12. Met Hounds Adlestrop House, Col. Arnold's – found in Oddington Ashes, ran out as if for Gawcombe, turned back thro' the Ashes past Adlestrop Stan. over the Park, & up the hill to, & thro', the Gorse, away over Chastleton, round, & lost a mile beyond Little Compton – Mrs. Callander broke her arm, & Mrs. Witts fell.

Dec 31. The hardest frost during this Autumn, and Winter: but could plough.

One of the worst years on record for Agriculture – cold, late Spring – wet Haytime for 6 or 8 weeks – late, bad-yielding, Corn Harvest – but average crops roots – good plant young Clovers, and grass much of it, eatable till Xmas – Wheat Red 26s. to 28s. Blk. Oats 14s. to 16s. 40 lbs. B. Barley 24s. to 28s.

1889

Jan 5. Hard frost, with frequent Fog, which freezes on the Trees, and has the appearance of a slight snow on everything – no wind. Mr. Price, tea'd.

Jan 7. Mr. Matthews & Son, Fifield, tea'd: and I walked up with them and attended a meeting at Leafield School, where Mr. Matthews addressed the natives as to representing them on the County Council – Mr. Maddox in the Chair – rather made a hash of it; Mr. Wynne, Maddox junr., & others, explained the matter – altogether a very tame affair!!!

Grace, John Chas., & Robt. Wm., attended Dr. Hyde's Dance.

Jan 9. A mild thaw – frequent showers – with wind – ploughing been stopped 9 days!!! The Fogs froze on the Trees, and the *weight*, broke down several of the branches.

Jan 14. Sold 93 Sheep at Shipton Auction, realised £218.1s.6d. . . .

At 6.30 attended Mr. Matthews last meeting, as a Candidate for County Council – Mr. Samuda in Chair – a very satisfactory affair, considering all the Refuse of Shipton & Milton attd.[1] Mr. Harold Matthews, made the best speech.

Had an hour, with Mr. Brookes – Home at 12.15.

Jan 21. Met Hounds Heythrop: found br. Foxes, in Longwood; ran one towards Little Tew, & lost, by being chased by a sheep dog, the other Covers blank – Mr. Price & I returned home, and called at Mr. Blisses, Chadlington: saw the wreaths on Revd. Carpenter's grave –

Mr. Brassey unable to ride, thro' a fall at Tangley!!!

Jan 24. The County Council held their first meeting at the County Hall Oxford – 19 Aldermen were elected; am pleased to say Mr. Matthew Savidge, Sarsden, headed the list with 46 Councillors' votes!!! – out of 54 present!!!

Feb 10. Snowed all the day – 5 or 6 inches on level.

At 12.20 Midnight my poor Wife, Jane Waters Calvertt, died – she was born the 4th May 1832 – taken ill last May, with Heart Disease – Jaundice ensued – followed by Dropsy.

Feb 13. Snow showers till noon – buried my poor Wife 3 p.m. – her Grave is about 10 yards off the entrance, into South-West Porch, of Leafield

Church.

Fair from 12 to 4, when rain set in, with a rapid thaw!!! The Revd. H.M. Ward, & Mrs. Ward, attended Funeral: from West Kensington.

Feb 23. My poor Jane's Death in 'Stamford Mercury', and 'Oxford Journal', viz.
—Fairspear, Wychwood, Oxford – and formerly of Tothill Manor, Alford, Lincs – Feby. 11th ult. Jane Waters Calvertt, in her 57th year.

March 14. Mr. Price drove me to Mr. Sartoris, Abbotswood, nr. Stowe: we were shown round the House, Gardens, &c. – Mr. S. gave Mr. Price, a Hunter, which he rode to Shipton.

March 19. Parish Vestry, at Fairspear, Mr. Addy, Gregson & Hobbis attended, we 'named' 6 men, & Charlotte Williams, Overseers!!!

Had a week's splendid Dry weather – planted early Kidneys, Chelots[2] – Trimmed Laurels, 200 Faggots – 105a. Spring Corn Drilled: rain at night.

March 30. Cambridge, won the Boat Race.

Hounds met at Cornbury – Chartreuse – offered, &c. I have not been on a Horse, since 8th Feby!!!

Had a delightful week's weather – 2 Drills on 5 days.

April 4. The County Council met at Oxford on Monday – Earl Jersey who has recently returned from India, was elected Chairman, in his absence abroad – they start very much like Business.

April 9. The Banbury, point to point, Steeplechases –

John Chas. rode Mr. Blisses Horse, 13 starters, he came in 3rd: which was a very *good performance*, in such *Company*, and a *strange* country.

April 13. The Duchess of Cambridge, buried at Kew – the Queen attended.

Mr. Price 1½ hour walking up here – all his life, been a great rider in Winter, and walker in Summer: *now* either, *wearies* him!!! in his 69th year!!!

April 26. Shipton Auction – Mr. Price sadly worse at Riding & Walking. I fear his Heart, is sadly wrong, & weekly worse!! . . .

May 4. Jane would have been 57 to-day.

From 5 a.m. to 8 p.m.: about the *finest* day of the *year.*

Mr. Price to Tea – Mr. Groves called one hour.

May 7. A *Terriffic (1½ hour's)* Thunder rain – washed thro' 7 of our Walls – been much hindered since 2nd April – it is 53 years since so heavy a rain fell around here.

May 9. Rode to Witney – Tea'd Mr. Reade's, Shipton, and accd. Mrs. Jenkins to our annual Primrose meeting in Beaconsfield Hall.

May 12. Miserable weather, cannot finish Spring-Corn sowing – Sow small seeds, Thrash, or clean land.
Spent the aftn. with Mr. Price.

May 22. Mr. Price rode with us – Grace, John Chas., & self – to Agl. Show, in Blenheim Park – George Chas. 8th Duke of Marlborough, and his Bride – Presided at Lunch, and appeared to thoroughly enjoy the whole affair – weather extra warm: the Palace, Orchids, Gardens, & Park, looked lovely.

May 27. 34 years this morning, since Alice was born, really what changes I have witnessed, in those many fleeting years.
Lord Churchill is cutting down all the Oaks, & 'smooth wood', next Cockshot-hill – for replanting with Larch.

May 28. Great alterations are taking place on Lord Churchill's Estate – part sold – much planted – and altogether much altered & improved!!!
The 'Earth-stoppers' Dinner at Charlbury – the report was *scarcity* [of] Foxes, to make more space for the magnificent bird, called, the Pheasant!!!

May 31. Much rain, and showery weather, since the *2nd April* – altogether stopped entirely, much work on the land, and none clean for Turnip drilling.

June 1. Rode round by Ascott, on to Ranger's Lodge, saw Mr. Harrod, who is rearing 2000 phts., then I rode past Cornbury House, to the Gardener's, which are 4 separate acres, enclosed by brick walls, the Orchard acre contains 50 Hen Pheasants & 5 Cks. – egg laying – their wings shortened for the laying season, after which they go into the Forest.

June 2. Never seen so much Grass, Clovers, & Vegetation at this date, during the 13 years I have lived in Oxon.
Mr. Price, drove with Grace, Isabel, & myself: to dine with R. Eagle, Stanton Harcourt: the Revd. Walsh joined us, he had returned from visiting his niece at Florence on the Friday, said he travelled from Milan to Oxford in 36 hours!!! showed me several Photo's, I was struck with the 'Freshness' of the White Marble leaning Tower of Pisa, 90 feet high, & 13 feet out of the perpendicular – Tradition says the architect built it purposely so, to show his ability!!!

June 3. Rode & called on Mr. Harrod, Rangers Lodge: and Mr. Temple, Cornbury Gardens: who gave me some Pansies for Isabel to sell at Leafield Bazaar to-morrow – 7s.9d.

June 4. Mr. Temple drove his daughter here, & looked round the Garden &c – left Flowers for Isabel at the Bazaar, as he retd. thro' Leafield.

Mr., Mrs., & 2 Miss Gasks – drove here and attended the Bazaar, which passed off very satisfactory, in brilliant weather.

June 9. After a week's threatening; rain set in, at 7 a.m. – for 4 hours. Whit Sunday. A great English Holiday, week!!!

June 19. Sworn in Churchwarden for Wychwood at Witney.

Gregson & Price, in evening.

June 22. Royal Show commd. in Windsor Park – covers 130 acres – this is its Jubilee year – and will be the largest and most successful meeting ever held.

June 25. Left Ascott 8.18 reached Windsor 12.10 – walked into the Castle, saw the State apartments, Jubilee Presents, St. George's Chapel &c – on the Show ground 1.10 – saw the Prince, Princess, Albert Victor, & his 3 sisters, of Wales. . . .

A grand exhibition of Animals & Implements –

Lunch with Messrs. Suttons.

June 30. Rode to Stowe – dined with Mr. L.H. Barker[3], who has . . . bot. the Victoria Brewery, £4500, of Mr. Tasker.

July 2. Two new knights – Sir Jacob Wilson – and Sir Robt. Nigel Fitzhardinge Kingscote, our Com[missione]r, Woods &c.

July 3. Carried Clovers 9 days, & Grass Mower cut 15 days, & *finished* 115a.

The Shah, or King of Persia, now in London – his 3rd visit to England, he was born in 1820 – Buckingham Palace is his head quarters, during his 3 weeks stay with us.

July 4. Miss Ada Lowe, & Grace, accd. me to Witney.

Mr. Alfred Groves & Sister tea'd, tennis, &c.

Mr. H. Gregson walked round Sheep, Beast, & Crops. . . .

July 14. 42 years to-day, since my greatest Friend, Mr. Barker, Claythorpe Hall, Alford, Linc: drove me to reside two years with Mr. Richd. B. Adams, Collow Grange, nr. Wragby, Linc: Oh! dear, what extraordinary changes amongst Farming Friends, & other acquaintances, in those fleeting years – shows the 'ups and downs' of Life – in all its ever changing scenes.

Mr. Smith, drove over from Filkin Downs, and Dined with me – he hails from East Linc.

July 22. Rode over to, The Place, Icomb – and farmed round with Mr. Saml. M. Simpson – 600a. walked to top of Tower and viewed the Country around – admired the 'Old Place', which is said to be the oldest House in Gloucestershire. . . .

July 26. Grace drove me to Cornbury Park – to a large annual Primrose gathering – Mr. Hinde addressed the meeting, after Tea-ing in the Riding School – Cricket and Dancing followed.

July 29. Magnificent forcing weather – sunny & warm. Hauling manure on Clover stubbles – Horse-hoeing Turnips – & stirring Fallows – Thatching Oat straw – Horse-raking Hay up in *Home Lawn* – about 10a. of it.

July 31. Mr. A. Groves came, and looked over the Kingstanding ½ Barn roof slating – and I accd. him to look over £1000 worth of works, they are doing at Heythrop House – Mrs. Brown showed me over much of the House – Mr. Jones, Butler.

Aug 3. Mr. John Young's Farm, 140a., of Shipton – bot. by Mr. Lay, Woodstock, £3000. Re-bot. for £3100 by Dangerfield!!!

Mr. Alfred Groves, & Sister Ellen – here in the evening.

Aug 6. Rode over to Hailey Manor, found Mr. Adams, sadly misfigured by a fall from his Horse – retd. by Cornbury Park, and joined the Revd. W.T. Lee and our Temperance Society – sent up 2 Ballons as a finish to a very pleasant afternoon's enjoyment for all the party. . . .

Aug 16. *All* hands, began tieing Wheats – 70a. cut.

Rode to Mr. Eagle's Stanton Harcourt, & *hired* 6 sh[earlin]g Oxfd. D. Rams.

Called at Mr. Adams Hailey, on returning, found him better from wounds caused by his Horse falling – rain at 5 p.m.!!!

Aug 31. Finished cutting 232a. Wheat: 30a. Peas & 75a. Oats already cut: and 195a. Oats & Barley to cut – 95a. Wheat & 30a. Peas, already carried!!! . . .

Sept 2. The 35 min. rain, of 7th May: drenched many of the Partridges off their nests – and drowned *many* of the *very young* Rabbits, in their nests – the Partridge shooting will not be nearly so good as expected in the localities caught with the heavy, extra heavy rains.

Sept 17. . . . Grand weather, comparatively secured Harvest – only 35a. Blk. Oats, left out: carried *Eleven* consecutive fine days!!!

Sept 18. Mr. Yapp walked with me over Cockshot-hill, & bot. 10 Sheep 50s. – Sent Mr. Mrs. & Baby Adams, Br. b[ir]ds – couple Landrail, & *Baby* Partridge.

John Chas., Grace and Miss Eliza Mason – attended an invitation Dance in Witney Corn Exchange – Mr. Smith, Filkins, met his Belle – Miss E.M. at the Marlboro'.

Sept 19. Rain set in, at 11 a.m. – and fell gently all the aftn.

Stowe Fair sadly marred by the rain.

Sept 30. Been a nice, genial, dry, growing month – for everything. An *immence* deal of Agl. Labour, done, this month.

Oct 23. John Chas. & Grace, met Hounds 'cubbing', at Tangley – killed brace.

Wool worth 24s. here & 25s. to 27s. in Linc. Alpacas are to be the wear – hence the rise.

Oct 24. Drove Witney – sold grinding Barley 23s. – Blk. Oats 16s.6d. to Birdseye!!!

Mr. Gregson walked in, & settled for Straw – Vetches.

4 Elephant Potatoes every eye cut separately – produced over 6 Bushels out of Mr. Wm. Shillingford's garden at Eynsham.

Mr. Wm. Clinch, Witney: bot. the 'Blanket Hall Brewery' with all its belongings, & added all to his 'Eagle Brewery'.

Stowe Fair.

Oct 28. Met Hounds, Sarsgrove – found a big dark Fox, and ran him to Chadlington, Dean, and crossed the Evenlode into Cornbury Forest – I left for Home – they found in Churchill Heath, and ran the Cub down, & went Home.

I called Mr. Adams, Hailey Manor – he had gone to Brighton for a week.

Rode down to Shipton Post Office for £7.15s.0d. retd. Inc. Tax. – wined with Mr. Price.

Oct 29. Attended Voly. School meeting at Revd. Lee's, 2½d. in £ rate. Shaylor called about wiring rabbits off the land next Forest.

Witnessed J. Edmonds perform with his Oxy-Hydrogen Lamp in 'Beaconsfield Hall'.

Oct 31. At Witney market. October, has been favourable for all sorts of work, and very mild throughout, much foliage is left on the Trees, and Grass, Clovers, *green* as May month.

Nov 1. . . . Attended a Parish meeting to hear appeals against the present rateable value, for laying a new County Rate – no objection from

Wychwood, but Leafield appeal, being assessed at £86 beyond their present rateable value.

Showers, wind, & sun.

Nov 4. Accd. Grace, to opening meet, at Heythrop – tasted Boar's Head but too rich for a mild morning – and *unsuitable* for Champagne, so tried Pheasant!! Found 4 Foxes in 'New Cover' ran one to Lidstone, & back into Cover, & killed – tried Walk Gorse, blank – I left. . . .

Nov 28. Our Primrose Ball at Beaconsfield Hall –

Examined the part of Forest being cut down all along our Cockshot-hill, and Witley Hill.

Too hard to plough.

Nov 30. Met Hounds at Eynsham Hall – Champagne & Pheasant with Mr. Mason – Found 1st Fox in 12 acres Gorse – lost him in Mr. Maule's yard Witney – found 2nd in Coggs & killed amongst felled Oaks in Mr. Townsend's Cover –

Called at Mr. Gask's on my way Home.

Dec 3. Walked to Mr. Temple's, Cornbury Gardens, he gave me some thoroughly black Potatoes, inside & out!!! Zulus!!!

Very *cold*, Foggy, misty day!!!

Mr. Price walked up, & lost 3 games Bezique.

Dec 13. Walked, and had Lunch, with Mr. Wynne's shooters, Lords Hood, Parker, Fitzroy – Col. Ellison – Major Norris, Capt. Walker & Sir Oswald Moseley. I shot br. ck. Pheasants & rabbit with Capt. Walker's 2nd. gun – 2596 head, were shot in *three* days, viz 442 – 1308 – & 866!!!

Dec 19. Many Woodpigeons in this locality – supposed many of them are from Scotland.

Yesterday, a dead man – found in Cornbury Forest – only 20 yards from the Finstock highway – supposed to have been dead 8 to 10 weeks!!! Job Cork, Abingdon.

Dec 20. Rode to Mr. Lowther Barker's, Stowe on the Wold – joined Mr. Brassey & Hounds – left them in *snow* and *rain*, at Eyford. . . .

Dec 31. The past year has been extra Productive in Wheats, Oats, Hops, Clovers, & Turnips!!!

1890

Jan 2. Fog & Hard Frost – the Thermometer down to 20 degrees!!! John Chas., Grace, & Isabel: attended an amateur performance in Witney Corn Exchange.

Jan 12. . . . Leafield Church was consecrated by Bishop Wilberforce on 19 Oct 1860 – The Lords of the Treasury gave £1000 – it cost £5000 – £200 of Queen Anne's Bounty[1] was given towards the Vicarage House.

Jan 24. Lord Churchill is replanting Cornbury Forest, with Oak, Sycamores, Larch, Spruce, & Scotch Firs.

Two men count & sort – two boys carry out – Four men place Trees in the already dug holes, and Twenty men, plant, & *finish* all nicely up.

Jan 25. Met Hounds at Sturdy's Castle . . . at 2 p.m. I left them, in wind & rain, trying Sq. Hall's Covers at Barton!!!

Yesterday had the best run of the Season from Eyford – 15 miles in 1 hour & 30 min!!!

Jan 31. A mild month – for Agriculture.

Feb 3. Wrote Will Dale, the Brocklesby Huntsman, to say Arthur Hazelton's, Huntsman's place, at Heythrop, was at Liberty!!! Drink!!!

Feb 4. Tea'd with Mr. Price.

The *two* pieces of Forest, at North end of Cockshot-hill, and *each side* the *Public Highway* from Leafield to Charlbury, are already planted with Oak, Sycamore, Larch, Scotch and Spruce Firs – the work is done *most satisfactory*; the *continued recent mild* weather, has exactly suited!!!

Feb 10. Shipton Auction – Mr. Wm. Mace bot. one of my Fat cull Ewes 60s.!!!

Feb 17. John Chas. & Grace, met Hounds at Chapel House, and found in Salford Osiers, and killed in Whichford Wood – &c.

Called on Mr. Mace, Langley: and saw an Ewe, with Thos. Hanks, which was sold at Shipton Auction, and claimed by W. Mace!!!

Feb 18. Met O.B. Hounds at Tar Wood . . . I retd. with Mr. Neild & Pain, had wine with Mr. Pain at Finstock.

Mr. Price rode to my House after I had gone – but so short of breath retd. to Shipton.

Four Policemen making further Enquiries about Mr. Mace's lost Sheep!!!

Feb 20. Rode to Witney – & explained publicly at the Fleece and other places, that I had had two Titles added to my name, since last they saw me, which were, that I was a 'Thief' & 'Sheep Stealer', to the dismay of my Friends!!!

Mr. G. Gask told me, he was about taking 'Bridewell' [Farm]. . . .

Feb 22. Called Mr. Brookeses: attended annual Beaconsfield Hall meeting, Mr. Brassey called 6 p.m. on his way Home from Hunting and took the Chair.

Grace met the Hounds, Shipton Barrow, killed, had abundance running.

Feb 24. Mrs. Price died yesterday, at Torquay – ae 64. . . .

Feb 28. Rode Mr. Price's Horse 'Belfast', and met Heythrop Hounds at Farmington – Frosty – ran & lost brace Foxes over Cold Aston &c – I left them trying the Grove – the Honble. Thos. A. Brassey told me, it was his *first* anniversary day of his marriage – with Lady Adelina Nevill!!! . . .

February has been, on the whole, a fine mild month, throughout!!! . . .

Influenza, is all the rage!!!

March 3. The hardest frost of the Winter.

Rode over to Mr. Gask's, who gave me all his Linc. news.

Mr. Price retd. from Torquay, after burying Mrs. Price in the Cemetary, on Friday last.

March 4. The Prince Wales, opened the new Forth Bridge, to-day: 8,000,000 Rivets were used in its Construction, and it may be said, to be, the '8th Wonder of the World'.

March 8. Rode and called at Mr. Gask's Bridewell, & met the Hounds at Blenheim – spent 4¾ hours in trying covers blank – 12 to 4.45 – when we found nr. Kiddington Gate – I left them 5.20 – near Wootton Church.

Had Whiskey & Cake, at Cornbury.

March 13. Wrote Mr. Mace about Falkner, the Potters Hill Shepherd, abusing me, at Shipton Auction, on behalf of Mr. Wm. Mace of Langley!!!

March 15. Met Hounds Cornbury Park – killed – ran 3 others.

Mr. Brassey ill from Influenza.

Only 6 Ewes left to Lamb – or Yean!!!

March 21. Spring quarter day – nice & mild – every present prospect of an Abundant Summer!!

Grace and I, met Hounds, Mr. C. Hobbses, Bampton Manor, found in Lew Gorse, ran 1½ hours, & killed at Blk. Bourton – Hounds went on to Casswell – we retd..

Attended Vestry Parish Meeting, Addy, Horne, &c.

March 31. Farmed round Water Eaton – with Mr. Henry Bulford – Messrs. Parrott & Smith, joined us at Dinner – heard a fearful account of the extravagance Duke Marlboro', and the doings of his Duchess (Hammersley).

The first *3 months* of this year, have been mild & dry – Water scarce already around here.

April 5. Met Hounds, Shipton: Mr. Mrs. Brassey, 3 Daughters, & 2 Sons, out: Hazelton's *last* day as *Huntsman*, after 22 years service with the Heythrop Hounds!!! . . .

April 10. By rail to Moreton in the Marsh Steeple Chases, had lunch with Albert Brassey Esqre. – a very good day's sport

April 16. Mr. Goschen's Budget – very popular – given away 3½ millions!!! and raised a Million on Alcohol & Beer, for the County Councils, & Compensation for Licences.

April 22. Mr. F. Matthews junr. in the evening – Grace & Isabel to Witney. Inspected the 200,000 recently planted Oak, Sycamore, Scotch, Spruce, and Larch Firs in Forest next Cockshot-hill – all appear to be alive.

April 26. A Lamb's throat cut in Kingswood, placed in a Sack bag, and dropped ¼ mile away – disturbed by the appearance of some one.

Tea'd Mr. A.G. Pain, Finstock: & saw the Policemen, & Keepers.

April 28. Mr. John Coles, Witney, died in his 50th year.

Mr. Clinch, Witney; and Mr. Abraham, Minster: both most seriously ill.

May 31. Leafield Feast yesterday, and to-day. . . .

May month, has been the 5th *comparatively* dry *one*, of the present year!!! the Country generally is looking remarkably well.

June 1. Walked down to Shipton in the aftn.

Mr. Brassey allows his friends to visit his Heythrop Gardens every Monday aftn.

June 2. Rode to Charlbury Auction.

150 of us dined with Mr. L.M. Wynne in Leafield School House – Revd.

W.T. Lee in the Chair – all passed off satisfactory – and most probably all present will vote for Mr. Wynne, next election!!!

June 13. Married 36 years ago as this day.

At Witney – had my Wife's ring made to fit myself!!! looked over Witney Cemetary – chose a grave stone for my poor Jane.

June 16. John Chas. drove me to Witney, to attend Bishop Stubbs Primary Visitation – appointed Churchwarden for 2nd year, parish of Wychwood – was Churchwarden the *24 years* I resided at Tothill Manor, Alford, Lincs.

June 19. Mr. Price accd. me to Ascot races – 'Gold', won the Gold Cup, he belongs Prince Soltykof.

Mr. C. Robinson the Judge, gave me a Paddock Ticket – had Lunch with Wynne – altogether one of the *very largest meetings*, ever seen, on Ascot Royal Heath.

June 20. Very busy time of year now sets in – all over the Country, Clover & Meadow, Cutting, Making, & Carrying – Turnip drilling, & cleaning, & singling – Fallows Dragging – Corn weeding.

Her Majesty began her Reign, *this day, 53* years ago!!! – 1837.

June 21. Grace drove Miss Ward, to Chipg. Norton.

Isabel, tea'd 8 College Companions, from Shipton.

Mr. Richd. Eagle, Northmoor – bot. Cow: £23.

The Cricket 'Players of England', are giving the 'Australians', a *terrible beating*, at the Oval; Gunn made *228*!!!

June 30. Had showers, and 3 fine rains, during June: which has secured Turnips, and Abundance of everything. . . .

July 5. Strikes amongst all kinds of Labour of frequent occurrence, owing chiefly to briskness of Trade in general.[2]

July 7. The Policemen striking in London!! what next?

Great storms of wind & rain, in many parts of England.

July 16. Rode over to Mr. S.M. Simpson's, Icomb Place: walked up Icomb Tower – and saw his Waterwheel in valley.

July 19. Mr. Mrs. Ed. Bliss, Chadlington Manor, of a Son & heir, on 13th ult.

1 inch rain fell, during night!!!

The 'Rollright Stones', are now enclosed for better protection, against visitors' mutilations – they are great relics of Antiquity!!! . . .

July 28. Very rainy night – and very warm, sunny day!!!

Mr. Yapp drove Mr. Cobb up here, and we walked down and inspected the Water Wheel.

Stovin made his first call here, as *new* Huntsman, at Heythrop Kennels.

Tea'd Mr. Price, & saw Mr. Pierce about Fredc.

July 29. Shipton Fete – John Chas., Grace, Isabel, with F. Gask, and Ada – tea'd at Mr. Price's, and witnessed the Sports &c. . . .

July 31. John Chas. drove me to Witney – tea'd, & looked round the New Bank, & House, with Mrs. Davenport.

Fredc. Geo. retd. from Burford School.[3]

July has been very growing, rainy, and in many parts of England, rain-stormy month.

Only completed Clover carrying, to-day!!!

Aug 2. By 12.58 for Newton Abbott – 12000 inhabitants – 6 miles from Torquay – stayed night at Bristol – a Temperance Hotel – very busy, dirty, dusty, smoky place.

Aug 3. Went on to Exeter – very picturesque finely situated town.

Aug 4. Reached Newton Abbott – a beautiful railway ride from Exeter, saw Exmouth, Teignmouth, Dawlish, &c.

Aug 5. By rail to Torquay, were driven around Torre, Torquay, & Babbicombe Bay – saw Daddy's Hole – Torbay – &c – returned to Newton Races. Mr. Blyth's 'Catspaw' won the last race – sold for 130 G. to Mr. Albury of Reading.

Saw egg, & bottle shooting, on the water: 'hand swings', &c.

Spent an hour with Mr. Mrs. Blyth.

Aug 6. By rail to Bovey Tracey, met Mr. Mrs. Huckvale, by Charabancs to Haytor Rock, on Dartmoor, Beckky Falls, Widdicombe Church, &c.

Aug 7. By rail to Totnes – down the River Dart to Dartmouth – saw Herons, Ducks, & 2 Salmon, caught by the *Totnes Bridge* – Lunch at Castle Hotel, Dartmouth, & left by rail for Paignton, where we attended a Concert at Mr. Singer's – Wombwell's Menagerie there.

Aug 8. By rail to Plymouth – went out in a Pinnace, and saw 17 English men of war, 2 American, & 1 French, all in the Sound: viz The Northumberland – Anson – Invincible – Inflexible – Triumph – Hotspur – Black Prince – Hero – Shannon – Basilisk – Inconstant – & other Gunboats.

Walked all round the Citadel – by train to Devonport Dock-yard – shewn all round – very interesting – the largest chain links weigh 3 cwt!!! 4764

hands at work, on the 75 acres.
Met H. Blyth in the evening.

Aug 9. By 9.44 train to Exeter 10.25, Bristol 12.4, Oxford 4.20, Ascott 5.5: after one of the *most various* week's amusement, I ever witnessed in my life – Hay carrying – and Corn Cutting, from Newton to Oxford!!!

Aug 10. A tremendous Strike in Wales – a *great Test Case* – of Labour against Capital: which will be watched by all classes in England!!!
Evening with Mr. Price.

Aug 16. . . . Began Pea Cutting in Kingswood.
Harvest ready all round here – only fine, fair Weather, required to secure it.

Aug 19. Rain!!! could not begin Wheat cutting.
Tea'd & spent evening Mr. Price.

Aug 30. Much rain has fallen all over the Country the past fortnight.

Sept 8. Carrying Oats 10 Wagns. Grand Weather – Showed Mr. Yapp Cornbury Gardens, &c.

Sept 18. Had Green Goose-berry wine, with Mr. Pain, Finstock.
L.M. Wynne Esqre. in residence at Cornbury.
Lord Brassey married Miss Capel for 2nd wife – and came direct to Wilcote House, for Honeymoon!!!

Sept 23. Rain & showers delaying the late Harvest, but grand weather – Finished Potatoes lifting – the best quality Magnums I ever grew.

Sept 25. Burford 'Mop' – only my 3 married Carters attended, the others prefer Witney. . . .

Sept 27. Carried Barley & Oats last 2 days – only about 16a. Oats to Carry – Delightful Autumn weather.
Mr. Price and Parsons, in evening.

Oct 1. . . . Leafield & Wychwood Harvest Festival: supped with Revd. T.W. Lee, and met Revd. & Mrs. York, Revd. & Mrs. Druce – Revd. Redifer – General & 2 Misses Rowcroft: &c.

Oct 2. Keen frost afterwards very *sunny*, with *blue* sky, the *exact weather* I like for *shooting* .
Mr. S.M. Simpson, Icomb Place – D. Westell, Witney – Dr. Hyde, Witney – Mr. Ed. Bliss, Chadlington – came & we shot 18½ br. Birds, 11

Pheasants, 10 Hares, & 8 Rabbits.
Mr. Price joined us at Dinner.

Oct 31. Mild drizly day, after another very Dry month.
On 23rd ult. – at Charmouth, Dorset; ae 71: Saml. Potter, of 36 King St. – Cheapside: shot with him for years at Strubby, Stenigot, &c[4] – a very intelligent man – really how many of my old, & even young acquaintances, are going over to the Majority!!!
Messrs. Wood & Clifton, have taken Withcall, nr. Louth, a Farm containing upwards of 2,500 acres: said to be the *largest single Farm*, in England!!! . . .
L.M. Wynne shooting in Cornbury Forest: next Witley Hill. Keepers keep beating my Land, & Hedges!!!
The Honble. Miss Dillon, going abroad for the Winter.

Nov 3. Grace accd. me, to Heythrop opening meet – we called on Mr. Mrs. Bliss, Chadlington.
Mr. Barnett proposed Mr. Mrs. Brassey's healths, said he had hunted with the Heythrop Hounds 54 years – so he had gained the title of 'Father of the Hunt' – &c – a very large concourse of People – all passed off very pleasantly after a magnificent Champagne Breakfast.

Nov 13. My 61st Birthday – J. Fras. Mason, sent me 2 Brace Pheas'ts.
Grace & I, attended a Primrose meeting at our Beaconsfield Hall – Miss Moore's young ladies amused the company very much – Isabel performed – Mrs. L.M. Wynne, drove over, from Cornbury.

Nov 18. Met Hounds 'Golden Cross', Ld. Willoughby cross and would have all the 'Field' together before he began to 'Draw'. . . .
Dined in Evening with Messrs. Price, Bliss, Maddox, & Groves, at Mr. Reade's.

Nov 30. Clear, keen, sharp, frosty morning, about 2 inches snow: *all* Old England, has experienced the slightest rain fall, the past year, for a very long period, especially Lincolnshire.

Dec 1. Frost, Fog, Drizzle, & cold N. Wind!!! . . .
The Parnellites quarrelling all round – every prospect of 'Home Rule' fading away for the next 10 years!!!

Dec 3. Mr. Mrs. Adams, drove from Hailey, & spent the evening – Mr. Dodson being sold up, by Mr. Clappen – on the *ruinous* Chase Woods Farm!!! under the Duke Marlboro'.

Dec 13. A week's still, keen, foggy Frost – extra *severe*, in Paris!!!
Mr. Parnell breaking up his Irish Party, to the injury of them *all*!!![5]

Dec 15. Walked down to Mr. Price's – so very Dark – came home Lantern in hand!!! Inky Dark!!!

Dec 25. My 3 sons, Grace, Isabel, & Miss Rose Smyth,[6] formed our Xmas Party – 6 to 8 inches snow, on the level, with still, keen, Frost – Trees all fringed with frozen Fog, & hoar Frost.

The Scotch Railways much injured by Strikes amongst their Employees. . . .

Dec 26. Had snow, frost, & thick Fog, for 4 consecutive days.

Walked down to Tea &c – with Mr. Price, he showed me Miss Pearson's Photo, or Carte – could not recognise it.

Dec 27. Read in the Standard an a/c of the Crown Land Revenues &c – nearly 70,000 acres of farm land are in Comrs. hands, and all 'let', but 11 to 1200a. – they also manage the New Forest 65,000 acres – the Forest of Dean 19,000 acres: Windsor Forest & Parks – they have 4252 Houses in London, assd. at £900,000: Rent £237,008 per year!!! The administrators of the Crown Property pay annually into the Exchequer £430,000.

Mr. Nathl. Clayton, died in St. Arches Church, Lincoln, on Sunday last in his 80th year: in 1845 Messrs. Shuttleworth & Clayton made their No. 1 Steam thrashing Engine of 8 horse power, and on Saturday last, their Firm finished No. 26,204 and 24,176 Thrashing Machines!!! Marvellous what 2 men can do in a Lifetime!!! Withcall was purchased in 1875 by Mr. Clayton.

Dec 31. John Chas. & Grace, gave a Dance, these attended, viz: Mr. Mrs. F. Habgood, Mr. Price, Mr. Akers, Worley, Simpson, 3 Matthews – Misses Matthews, Akers, 2 Gillett, Worley, Simpson, Smyth, and Mr. Mrs. Emerton, & L.H. Barker.

The past year has been one of the *driest* on record, rather over an average of all kinds [of] Agricultural Produce.

1891

Jan 1. Frost and snow, all over England, with *keen dry* air!!! . . . John Chas. cut his Great Toe with 'Billy Hook'.

Jan 2. Mrs. Adams sent us some Oysters.
Mr. Lane, Surgeon, came and saw John Chas. cut Toe. Grace and Rose Smyth, attended the 'Yellow Dwarf', at Beaconsfield Hall.

Jan 3. Mr. Price resolving to marry Miss Pearson.[1]
Grace, and Miss Rose Smyth, again attended the 'Yellow Dwarf' at Beaconsfield Hall – Mr. Wynne, & party, there.
L.M. Wynne finished up the Cornbury shooting for the Season and now going to Mentone.

Jan 4. A month this morning, the ground was covered with snow, and been with an inch, to 8 inches on the level, ever since, with every appearance of continuing.

Jan 10. Severe frost all the week – every appearance of one of the greatest Frosts, England has known, for a *great many* years.
Alice Mason, Eynsham Hall, married to-day to Capt. Longfield, at St. George's, Hanover Sq.

Jan 19. Walked over & saw Mr. Abraham, Ringwood: been a great invalid thro' Rheumatic Gout for a *year* – helpless in his legs & arms!!!

Jan 20. Mr. Price married Miss Gertrude Pearson, at Combe, in Hampshire, nr. Hungerford.
Walked down to Shipton, and saw Mr. Pierce about the Bells ringing. . . .

Jan 22. Rode with John Baughan, down to Mr. Price's, Shipton, he and his Bride reached Home from London, by the 5.10 train – very merry, and well.

Jan 24. *Seven* weeks to-day, since I hunted at Blenheim – the same time since the Cart horses ploughed, and they cannot plough to-day!!! Mild rain. . . .

Jan 25. Walked down, and Tea'd with Mr. Mrs. Price, the *first Sunday*, of their married lives!!!

Jan 30. A general thaw all over England.

Terrible collapse (for the men) of the great Scotch Railway strike.

Mr. Bradlaugh M.P. for Northampton, died – a singularly prominent Politician – he was the son of a solicitor's Clerk.[2]

Jan 31. Met Hounds, Kiddington Gate, tried Ditchley Park and Covers blank – found in Kingswood and ran up & down the Cover till we lost in a great squall – I left at 2 p.m. Hounds went on to Swan's Nest. . . .

Wheats, Clovers, Grasses, Turnips, not nearly so much affected after the severe Frost, as expected!!!

Feb 2. Shot Rabbits at Ranger's Lodge, with Mr. Wynne – Mr. Jacobs, & Cancellor, & 3 London gents, joined us.

Very nice day's sport shot Rabs, 6 Ck. Phts., & Jay – had Lunch in centre room, of Ranger's Lodge – for my first and last time, as I believe it will be taken down this Summer.

Mr. John Jacobs, tea'd, and spent evening, at Fairspear.

Feb 10. At North Leigh Church, Mr. Dean married Ada Gask.

Met Hounds at Burford, found in Jolly's Gorse and ran to ground in Brash Copse, after a 7 miles run – found another brace Foxes in Jolly's, and lost in Burford – called on Mr. Walker – hounds found again in Bradwell Grove and ran till dark – left off beyond Brize Norton – did not reach Home till 9 p.m.

Feb 18. John Chas. drove Grace, for Tea, Mr. Sutton's, Wilcote.

Planted early Kidney Potatoes.

Lambing slowly the past 21 days!!!

Plashing hedge around Langley Grove – it will lay *quiet* – for 12 or 15 years!!!

Feb 28. Altogether, Feb: has been a very mild, still, agreeable month, for all kinds of Labour, and Pleasure.

Met Hounds Kiddington – killed 1st Fox – lost 2nd and tried very much of the Country Blank – even Dean Grove – to finish the day.

March 7. Rain set in 6 p.m. – after *6 weeks* dry weather, since the *8 weeks* Storm broke up – will be most satisfactory to all the Country – as water is scarce in many villages!!!

March 9. Shipton Auction – 113 Ewes & tegs – sent down & sold – £239.10s.6d.

A terrible snow storm, walked thro' it, after midnight!!! the worst journey up here, I ever experienced.

'The Royal Sovereign' will be the heaviest ship in the World, displacing no less than 14,150 tons of Water, and drawing 27½ feet when fully equipped!!! her 4 great guns perched on her barbette Towers, weigh 67 Tons each, and consume 530 lbs. of Gun powder at each shot, will throw a bolt of 1250 lbs. weight thro' 26 inches of unwrought iron at a distance of 2000 yards – her secondary armament consists entirely of rapid firing guns, and 10 of them weighing 5 Tons each, can be made to hit a target at 1000 yards, with 100 lbs. Balls, 4 times out of 5, in a *minute*!!![3]

March 11. Mrs. A.W. Tyson, wrote me a doleful account, of her Health which I much regret – she is the *last* of the Ashby Calvertts.[4]

Digging out Snow, on the Highways and Byeways, is very fashionable just now.

March 21. Rode thro' Forest to Fawler Bridge in 40 minutes, and on to Black's Lodge, in another 40 minutes: and forward to the 'Bear' Woodstock – had Cherry Brandy, with Mr. Hawkins: before the Hounds reached the 'meet' – we tried the 'Woodstock Farm' Covers, Blaydon Wood, Begbroke Wood & Burleigh Wood, *blank*, found a wired Fox in Pinsley – ran ½ mile and killed – found again in White Hill Cover and lost in North Leigh – tried Wilcote blank and went Home – I called at Bridewell, and tea'd.

March 23. Mr. Mrs. Abbott & Son[5] – started from Bombay, in the P. & O. S.S. Peshawar on the 14th ult. expect to reach the London Docks in 24 days, and call at Aden, Suez, Port Said, Brindisi, Malta, & Gibraltar.

March 26. Rode [with] Mr. Matthews to Witney, tea'd Mr. Davenport's and returned here after Masonic Lodge, attd. by Mr. Matthews.

March 28. Hounds met Ascott – ran brace, and killed one Fox, from Stockley. . . .

Mr. Mrs. Price, and Miss Pearson, called, while hunting on Wheels. . . .

April 2. John Chas., Grace, Isabel, & Miss Rees, by rail, to Morton Steeple Chases.

John Chas. rode Mr. Blisses 'Sarchedon', and won, there were 7 chases – all passed off satisfactory.

April 5. The 10th Census, taken to-day.

Walked to Mr. Price's, and told him the week's Hunting news.

At Church & Sacrament, in a.m.

April 6. . . . Thirty nine years to-day, since I commenced Farming, under Lord Willoughby de Broke, at Tothill Manor, nr. Alford, Linc. Regret I did not go to *Law*, instead of the *Land*. . . .

April 9. Mrs. Abbott and her Son Bernard reached here, by 5.5 train, left Bombay on 14th March and reached India Docks in 25 days . . . Mr. Abbott and Miss Edith, remain in London yet.

April 12. Mrs. Abbott and Bernard went to Church, but returned before the Sermon – from Cold – last Sunday they were 'rolling' in the Bay of Biscay!!!

April 18. Grace, accd. me to Kiddington Gate – to meet Hounds, ran 3 Foxes upwards of an hour each, 1st to Earths, the other brace (Dogs) killed in good style, and finished the Season – had Sherry & Ale, and Bread & Cheese at Mr. Ponsonby's.

The worst season's sport of the past 17 years, of Mr. Brassey's mastership!!! Drought, Frost, Fog, & Snows.

April 30. Slight rain set in from the West at 5 a.m. – and continued more or less, *all* day; to the *great satisfaction* of every-thing and every-body.

Lord Allington's 'Common', won the 2000 G's. Stakes, at Newmarket, yesterday: betting 9 to 1 against.

April has been a very arid, cold month – Vegetation very backward – more so – than for several years.

Best Red Wheats up to 40s. to 42s. – around here – not been so high for years!!!

Mr. Mrs. Abbott and Bernard here: John Chas. in Linc., and Grace, in London, Isle Wight, & Southport.

Much of Ranger's Lodge, taken down.

Mr. Morrell beat Benson, for mid Oxfordshire, by 688 votes – Sir George Dashwood, seriously ill – from excitement – after over-exertion for Benson!!! . . .

May 1. Very showery day – all so glad – Water is even very scarce, in *many* Counties.

On April 23rd Mr. Edwin Dodson, Gunsmith: Louth, Lincs – died ae 75 – I bot. my first 3 or 4 new Guns of him – began with Wads, Caps, Shot and Powder – all separate, in my Pockets – *advanced* from the above *slow* process, to *Breechloaders*, and *'Dart's Eagle'*!!!

On April 29th 12000 Sheep penned at Louth: Hogs 45s. to 52s. *only*!!! Sheep not been so *cheap* for years, not paid scarcely anything for their *Winter's Keep*!!! Lincoln Fair's Hogs did not average 50s. each – 25,000 penned!!! 3 years old Bullocks, for grazing; only fetch £18 to £22!!!

May 6. Had 7 nice suitable growing days, which has much improved appearance, of all kinds of Vegetation.

Parish Ladies Tea.

May 7. Mrs. Abbott & Son accd. me to Witney.

Attended Vestry Meeting at Mr. Howse's – appointed Wm. Wilkins, Ramsden, Asst. Overseer for Wychwood, Salary £9 per year.[6]

John Chas. retd. out of Linc.

May 13. Mr. Henry Arthur Brassey, died at Preston Hall, nr. Maidstone, ae 50, only Brother to Lord Thos., & Mr. Albert Brasseys!!!

May 16. Mr. Abbott reached here from London, by 5.5. train: very unwell, from Malaria Fever, Ague, &c.

May 17. Rain, hail, and snow squalls, from *West*, Rime, & blk. Frost!!! I fear injury, to the 'Great Bloom', on Fruit trees. Rain & Snow most of the night.

May 18. Rain and snow, all day: quite spoiled the Cornbury Sports. Leonard, Harrild, Edgar, Edwin, and Misses Hilda, Evelyn, Gwendoline, Beatrice, and Ruby; are the names of the late Mr. Henry A. Brassey's, 4 Sons, and 5 Daughters!!! the funeral took place yesterday, at Aylesford.

May 21. Robt. Wm., Mrs. Abbott, and myself, were driven to Witney – we looked round the Cemetary, and chose a Gravestone for dear Jane. . . .

May 24. Queen Victoria's 72nd Birthday – she will Live and Die, the grandest Monarch, who ever sat on a Throne.

Rainy day – had rains, and showers, hail, & snow, since the 30th April – none too much Water yet, as we are much below the average rain-fall: for the past 2 years!!!

May 31. Never saw so much 'Bloom' on *all* kinds Fruit Trees, never saw it so late, or remain so long before 'setting' – the late sorts of Apple trees, are *not yet* in full bloom!!!

At Church – Mr. P.S. Yapp, in full residence.

May has been cold & wet – with slight sunshine.

June 4. Sir Wm. Gordon-Cumming Bart., accused of cheating at Cards while playing with the Prince of Wales, and party, at Tranby Court nr. Doncaster, on the 8 & 9 Sept ult. – the Game was 'Baccarat'. The case is causing much *sensation*, the Prince of Wales, being Banker at the time.

June 11. Mr. Abbott accd. me to Ascott Races – we saw 'Bumptious', win the first race – 'Common' so far the best 3 yrs. old of the Season, win the St. James Palace stakes – Marq. Huntingdon's 'Morian', win the Gold Cup – weather grand!!!

Saw Sq. Heneage, Visct. Valentia, Sq. Mason, Eynsham, had wine with Mr. Abbott, at L.M. Wynne's Coach – Mr. Mrs. Price accd. us.

June 15. Mr. Abbott, to Miss Pulford 53 Guildford Street, Russell Sq. & Isabel to Mrs. Palmer, Collingwood House, Leyland Road, Lee, Blackheath, London.[7]

June 27. . . . Frequent showers day & night, all the week, consequently *most* growing weather, Thermr. 60 to 75!!!

The Crown entering on Mr. Wm. Mace's Fallows at Langley.

June 30. The German Emperor & Empress reach Windsor on 4th and visit Her Majesty till the 14th inst!!!

June has been most favourable weather for Agriculture.

July 6. The Emperor Wm. 2nd reached Windsor safely on Saturday – he is very busy every *hour* of his time – it is marvellous what he will *witness* and *perform*, the next 5 weeks.

Mr. Mrs. Price drove up.

July 8. Rode over and farmed round with Mr. Simpson, Icomb Place.

Rain *spoiling* Clovers – & sadly affecting Turnips.

July 21. Mr. Algernon Abbott junr. came on a visit.[8] . . .

July 24. John Chas. drove Mrs. Abbott – & Mr. A. Abbott junr. to Fifield, to bid adieu to Mrs. McCullam (ńee Miss Matthews) before she sails to New York, and Toronto!!!

July 31. Frequent showers, and very little sunshine, during this month. Floods, & very cold in many parts of England.

Harvest, & Vegetation, backward, *all* the year.

Aug 3. Cornbury Fete – the Beckwith family did a lot of swimming in the Lake – the two Sisters have handsome features.

Mr. Dickson-Hartland, and Mr. Richards spoke.

Thunder showers sadly marred the great meeting. . . .

Aug 5. Mr. Wm. Clinch, Banker, Brewer, & Farmer, Witney: died ae 71.

Aug 7. Finished Clover carrying for this season – been a long tedious affair, much Fodder all over the Country injured by the continued showery, sunless weather.

Dr. & Mrs. Wilcox, & Roberts; Mr. Worley & Miss Akers, Mrs. Gask & Son, Mr. Hayward, and Miss Simpson, & Mr. Fredc. Smith, and Matthews: came for Tennis.

Aug 10. Mr. Blakeman Eagle, Manor Farm, Thrupp – drove me round Blenheim, called on Mr. Angus – saw their Horses, Beast, & Sheep: on the

Blaydon Farms: &c.

Mr. B. Eagle's Thrupp Farm contains 454 acres, and 152 acres of Glebe – The Farm belongs Exeter College.

Mr. Parrott & Bulford, joined us at Dinner.

The Canal, & Railway, pass close by the Manor House, 164 trains run by during every 24 hours!!!

Aug 12. Grace 23 – drove to Mr. Habgood's, Gigley Farm.

Mr. Fredc. Matthews came in evening, and invited Grace to accompany him to Birmingham, to buy an Engagement Ring.

Aug 13. John Chas. drove Mrs. Abbott, Fredc., & myself, to Witney.

Best Red Wheat 41s. bid to Mr. Franklin, Capps Lodge.

The late Mr. Henry Arthur Brassey's *personalty* over a Million!!! and the *poorest*? of the three Brothers. Lord Brassey's wife Miss Alnut, was a Millionairess!!!

Aug 21. A very great rain – the greatest of *this* year. . . .

'Gloucestershire' only made 25 in their 1st innings with 'Somersetshire', at Cheltenham!!!

Rode over, wined, & tea'd, with Mr. Adams of Hailey Manor – rain aftn. & evening!!!

Aug 25. Dull, miserable morning, after rain in the night – there is rather an extra Crop of everything in the Country, if it can only be secured in good Condition!!! which is ruinously doubtful at present.

Aug 29. Fearful, & ruinous, Winds & rains, over all the Country – immence Damage done, to *all* kinds of Fruit, & *Grain* Crops!!! . . .

Aug 31. Rain in the night – most annoying, cannot begin Harvest. Wrote Mr. Wm. Mace!!! to settle up at Michs. for the trespassing of Sheep and Beast, the past 16 years.[9] . . .

Sept 11. Wonderful!!! had 4 *whole*, fine days, for cutting Corn. Mr. Algernon Abbott junr., Bicycled here, from Crondall, Hampshire.

Cut 180a. Wheats – all of it – in 5½ days, with 2, and 3 Reapers – how expeditious, and what a saving of time!!!

Sept 14. Rain, rain, hinder, hinder.

Shrewsbury, Gunn, & Abel, the three Champion Cricketers of the Season; averaged over 48, 41, & 35.

Sept 19. Only carried Corn Wednesday and Thursday up to this date, rain, rain!!!

Sept 21. Rain, rain.
Mr. A. Wm. Abbott, became a Sub Lieutenant in H.M. Royal Navy.

Sept 23. The Agl. Show, at Chipg. Norton – poor affair I *heard.*

Sept 25. Called upon L.M. Wynne, Cornbury House, and offered him High Lodge portion of Farm, 538a. 2r. 14p.
Wrote Crown, *first* letter, with regard to *Quitting* at Michs. 1892.
Burford Hiring and Fair – always held on 25th Sept.

Oct 5. Extra Busy – carrying Farthings Wheat to Machine – for Seed. A Reaper cutting Oats in Big Field – and 3 Waggons carrying Oats out of Rickyard field – Glass falling!!!

Oct 6. Rain wind rain – all day!!!
Harrod called, with regard to my High Lodge shooting for L.M. Wynne Esqre.
Mr. Matthews' Stock &c – Sale, Fifield.
Shipton Harvest 'Thanksgiving'?? Corn out all over the Parish – drenched!!!
Charles Stewart Parnell, died ae 45, the Champion of Ireland. . . .

Oct 10. This day completes me 16 years occupation of Fairspear & High Lodge Farms!!!
Only carried Corn 8 days, during the last 4 *whole* weeks!!!
Rain, rain!!!

Oct 14. Rain and terrific wind in night – with Thunder, & *heavy* showers, during the day.
Wretched time for the late Harvest: 200 acres, Wheat 15a., Barley 23a., & 162a. Oats!!!
The Revd. W.T. Lee, & Mr. Lion called.
Mrs. Wilcox hears that Mr. L. Midgley will probably take South Lawn!!!

Oct 19. Miss M. Savidge, Shipton: married Revd. Upstone. . . .
Rain from midnight to noon!!! Ruinous!!!

Oct 22. Mr. Mrs. & Bernard Abbots left by 12.58 train for Brighton.
Rode to Witney – rain!!! ruin!!! all round!!!

Oct 28. Seven men *mowing* Oats in Kingswood, & *finished* cutting Harvest!!!

Oct 29 & 30. Carrying Oats and Wheat – after *24 dear* days delay!!!

Oct 31. Mr. P. Yapp left Loughboro' House, to spend the Winter at 200 Sloane Street: London.

October has been the most disastrous weather for Harvesting only carried Corn on 3rd, 5th, 30th & 31st!!!

Nov 2. Grand opening 'Meet', at Heythrop – Grace and Miss Minnie Rees, drove – abundance Foxes – small attendance of Hunting men, abundance of 'Eaters'!!!

Nov 7. This makes 8 Consecutive Corn-carrying days, and Finished!!!

Only began wheat, hand sowing!!!

Fell in Breakfast Room and injured my Spine . . . applied an Embrocation.

Nov 13. Mrs. A.E. Abbott sent me, an Aueroid Barometer from Brighton, as a 62nd Birthday Present.

John Chas. met Old Berkshire Hounds at Bampton – found in Lew Gorse – he jumped his Horse into Lew Ditch – fished his Stirrups out of it – and retd.!!!

Nov 17. Dined at Mr. Reade's, Shipton: met Messrs. Price, Bliss, Groves, and 2 Maddoxes – the Rent audit.

Sept, Oct, & Nov – to now: have had the most ruinous wet Weather for the Harvest, and Wheat sowing, I have ever experienced the past 39 years I have farmed. . . .

Nov 27. By 8.17 train, reached Paddington 10.25 met Mr. Abbott in Chancery Lane – called & saw G.E. Skerry at No. 67 – & Fredc. – arranged that he secured Fredc. a situation. Left 3.30 Ascott 7.10 – *Incident*, with a man in train.

Nov 28. Finished wheat sowing for *this* year – never had such difficulty in carrying Harvest, & Wheat sowing, the past 39 years!!! Thatched all – & Buried Turnips!!!

John Chas., met Hounds, at Eynsham Hall.

Nov 30. Very thick, foggy day — Hounds could not hunt.

Both the Misses Simpson, Icomb – came, and sang at Leafield Concert.

Wretched wet, windy weather – during Sept – Oct – and the present month!!!

Dec 2. Too unwell of Cold, to shoot at Ranger's Lodge.

Very rainy last night: extraordinary: this starts the 4th consecutive month, of Jupiter Pluvius!!

Dec 5. By 12.58 train to visit my old friend Mr. Colton Russell, 50 Chestnut Street, Southport: I travelled via Rugby, Nuneaton, Lichfield, Stafford, Crewe, Warrington, and Wigan.

Mr. Russell accd. me to Liverpool – we took Packet to Birkenhead, saw the American Bullock Lairs – saw them Slaughtered for Mkt. – &c – returned by rail *under* the Mersey – saw Liverpool Exchange – &c – it is one of the busiest cities in *Europe*!!!

Saw Othello performed by Mr. Mrs. Teale, Miss Maud Stewart was Desdemona.

Saw the late 'Lamented', Mr. Nicholson – performed, Mr. Mrs. Crosse, & Mr. Mrs. Webbe, 'Larkey'.

Altogether had a most enjoyable visit. . . .

Dec 14. Walked down to Shipton auction & spent evening with Mr. Mrs. Price, Sargent Franklin & Shillingford, there.

Mr. Wakefield, Eynsham: absconded with lots of money, and his Governess – left wife, and 4 children behind – !!!

Dec 17. . . . John Chas. & Grace gave a Dance to 32 – viz, Gilletts 4 Matthews 4 – Simpsons 3. Piggots, Emertons, Robertses, Wilcoxes, Gasks, Habgoods, Prices, Smith, Yelf, Barker, &c.

Frost set in.

Dec 23. Grace, disagreeable to Mrs. Abbott, to my great regret. Frost, hoar frost, ground hard, & still – suits exactly, for Hauling, thrashing, skating, & Xmas Eatables.

Dec 25. Mr. Mrs. & Bernard Abbott, Mr. F. Matthews, John Chas., Robt. Wm., Grace, Isabel, & myself, dined together.

Dec 26. A thaw set in 6 a.m. – wind followed by rain till midnight – when it cleared up and again Frost set in. Mr. Mrs. Abbott & John Chas. attended Mrs. Samuda's Concert in the Beaconsfield Hall, Shipton: Grace & Isabel stayed night at Shipton, with Mr. Mrs. Price.

Dec 30. Rain & Fog!!! mild – Wind West.

Mr. Wm. Shillingford paid 2s.6d. in £, in 1881: *£32,000 default, now* £4900 – £800 only in hand!!!

Dec 31. The past [year], has been the most wretched, ruinous year's weather, for farming, I ever experienced since 1852!!!

Jas. Francis Mason, sent me 2 brace Ck. Pheasants!!!

1892

Jan 5. Mr. Robert Hambidge, Ascott, died on 1st Jany: ae 64. Mr. Mrs. Abbott, John Chas. – Robt. Wm. – Grace, & Isabel, attended Witney 12th Night Dance: Mrs. Price, & Fredc. Matthews junr., accd. them.

Jan 6. Influenza – alarmingly severe!!!

Jan 13. Can only Thrash & Haul – an inch snow – hard Frost.

The Influenza generally prevalent – mainly aged persons die – there are 4000 Cases at Bury St. Edmunds.

Jan 14. The Duke Clarence said to be in a critical condition, with Influenza, & Pneumonia.

Mr. Mrs. Price, attacked with Influenza.

Mr. Mrs. Abbott, were driven to Witney. I am too unwell, from Cold, to accompany them.

Cardinal Manning, died at 8 a.m. – ae 84 years.

Duke of Clarence died at 9.15 to the *great regret* of Millions of Her Majesty's Subjects – ae 28 years.[1]

Jan 15. Mr. Mrs. Abbott – John Chas., Robt. Wm., & Isabel – spent the evening at Mr. Mrs. Simpson's, Icomb.

Persistent dry Frost, with an inch Snow – day by day.

Duke of Clarence death has caused considerable consternation, the world over; 28 years old on 8th inst. and was to marry Princess Mary of Teck on 27th Feby. inst. – has turned a joyous Family into terrible mourning!!!

Jan 27. . . . Grace been the night with Mr. Mrs. Price – Mr. Price very ill from Bronchitis.

Jan 29. Mr. Wm. Wells, Charlbury, called & bot. 25 Fat Ewes, 45s. only!!!

Mild windy day – John Chas. to Icomb.

Mr. Dean of 12 acres very ill of Sore Throat & Lungs also affected.

Mr. Mrs. Fred Habgood, recovering from this wretched Influenza.

Everyone in Heythrop House, had the Influenza, *except* Mr. Mrs. Brassey!!!

Mrs. Adams, Hailey Manor, recovering from Influenza – when taken you ache all over, shiver with Cold, then a biting fever sets in, no appetite – lie in bed 2 or 3 days, & eat and drink, everything Hot, and as much of it as you possibly can.

Jan 31. Walked down to Mr. Price's found him very weak and much wasted during the past week!!! – I despair of his recovery!!!

Feb 1. Attended a Directors meeting in Beaconsfield Hall Shipton, Mr. Joseph Reade junr. in Chair.

Spent 2 hours, with Mr. Brookes – Mr. Franklin called: while there, heard Mr. Chas. Haddon Spurgeon died last night at Mentone: ae 58 – a remarkable preacher.

Feb 4. At Witney – awful bad market Wheats down to 30s. Barley 23s. Oats 16s. Wool 21s. Mutton & Beef 2s.4d. to 4s.6d. only!!! Monday Tuesday and to-day finished hand sowing and Dragging in, 54a Wheat on Kingswood.

Feb 11. Spent the evening with Mr. Mrs. Reade – heard a very *sad* description of my best neighbour Mr. Price's affliction, I much fear, he will never go here again!!!

Feb 12. Received a most important Document from the Crown offering me Fairspear Farm 660 acres, £600!!!

Feb 14. Extra fine, mild, still day. Afternoon, at Mr. Price's – Dr. Cheetle called – sad account of Mr. Price's legs enlarging!!!

Old Poacher Longshaw, Shipton, shot off part of his Foot, Dr. Cheetle took off the remainder – been the *greatest* Poacher, in Shipton, for many, many years!!! ae 63.

Feb 19. Snow, a cutting East Wind, make matters miserable!!! Called on Mr. Price, legs swollen – otherwise better??

Miss Mary Simpson, and Mr. Wm. Matthews came by rail, & attended, & sang, at a Concert at Leafield: weather *severely cold*, & wretched!!!

Feb 21. A thaw!!! for which I am very thankful.

Had Champagne with Mr. Price, who I found rather better. . . .

Feb 26. Rode and called on Mr. Price, then forward and called upon Mr. Mrs. Keck, Bourton on the Water: and met Hounds, found a Fox in Clapton Gorse, ran thro' Sherborne Cow Pasture and on to Sherborne village, back into covers by the Park wall and killed in 45 min – very good run – found 2nd Fox at Turk Dean & lost in Cold Aston village – refreshed at Mr. Acock's – tried his Lordship's Gorse, and Stow bridge Copse, blank!!! refreshed at Mr. Butler's, Wyck Hill, and at Mr. Matthews, Fifield. . . .

Feb 27. Rode over and call[ed] on Mr. Mrs. Bliss, Chadlington: found him in Bed, but much better from Liver &c – also called on Harrod, at Ranger's Lodge.

March 3. Mr. Matthews tea'd on his way from Witney.

Recd. Mem'm. from Crown Office offering us Fairspear Farm & 7 Cottages, on annual Tenancy, Rent £450, 659a. 3r. 17p., we pay all outgoings, except Tithes, & Income Tax.

March 26. Received Col. Sir Nigel Kingscote K.C.B.'s acceptance of my offer, of £450 per annum for Fairspear Farm of 660 acres, including 7 cottages: and Tithe free. . . .

On 16th Fredc. George trained to Liverpool, and next morning went on board the 'Parisien', for Halifax, on his way hence, by rail, to Winnipeg.

March 31. March has been first half – hard Frost: the second half, not so severe, but constant, & rimy!!! consequently, all kinds of Vegetation, very backward.

April 2. Called upon Mr. Bliss, Chadlington, and met Hounds at Ditchley – did not see the *new* Visct.[2] – regretted to hear that he was no Horseman!! he has only *one* Son, ae 18 – Covers blank – extraordinary – we only found *one* Fox all day, altho' 4 Foxes were found in my Ascott Common & Big Brazil *yesterday*, while 8 of the Keepers were looking for Plovers' eggs!!! they did not find an egg – the Hounds tried several of the Ditchley Covers, much of Cornbury Forest, & Blindwell!!!

April 4. Mr. Lenthal Kyffin met the Old Berkshire Hounds at Wantage during the day, while riding thro' a Farm Yard was seized with an apoplectic fit, fell from his Horse and instantly died, to the great regret of the whole 'Field', by whom he was much respected – ae 69.

April 9. Met Hounds, Shipton Barrow, only found one Fox in 'Stockley' ran him till we killed – tried all the Covers blank, till they reached my 'Boynal', ran hence to Priest Grove, Langley Groves, Egg Barn Cover, & to ground in Mr. Price's Drain, so ended a very irregular, moderate Hunting Season!!!

Capt. Middleton, the daring rider across Country for many years, was pitched off his Horse & neck broken, while riding one of the Steeple Chases at Kineton.

April 21. Robt. Wm., drove me to Witney.

Appointed Churchwarden for Wychwood.

Mrs. Abbott & Bernard came to spend a few days.

April 30. This has been a cold, rough, Dry month: vegetation has scarcely started yet!!!

May 2. Mr. Peter Yapp, taken up his residence at Lowborough House, Leafield, for the Summer – from 200 Sloane St. London. . . .

May 17. Rode to Fifield, assisted Mr. Colcutt, the Architect to the London Indian Institute – & Mr. F. Matthews junr., to shoot 106 Rooks, my Schultz's Cartridges killed them at great distances.

Mrs. Colcutt, Mrs. and Miss Matthews tea'd at Fairspear.

May 30. Accompanied Grace and Miss Hanger, to Mr. H. Matthews of Taynton, and spent a very pleasant evening, Mr. M. told me much, & shewed plans of Vancouver, &c – where he has purchased various Properties: in Canada.

May 31. Some fine showers fell, all round the Counties – Thunder. This has been *another* very Dry, Cold Month.

June 10. Dined at Leafield 'Potter's Arms' – Club Day.

Tea'd and spent 2 pleasant hours with Mr. Yapp & Sons: at Lowborough House.

June 21. Drove to Chipg. Norton Junction, by 8.17 train via Banbury, to Warwick Royal Agricultural Show – joined Mr. Simpson & Knowles. Saw the parade [of] Beast at 11 a.m. – The Prince Wales, Duke of York, Earl Warwick, & Countess – Lord & Lady Brooke & the Honble. Guy Greville – Prince Christian &c &c in the stand.

At 12.30 attended Members' meeting[3] – Ld. Feversham in Chair, Lord Leigh, Lord Brooke, Prince Christian, Duke Westminster, Marquis Hertford, Earl Lathom, Pole Gell, Sir Nigel Kingscote, & others addressed the meeting.

The most magnificent sight I ever beheld at a Royal Agricultural Show – Warwick Castle, Church, and the Avon, added much to the scene.

Dined with Mr. Simpson & Knowles, at the 'Crown', after which we walked in the Gardens – saw the 'Maze' &c.

June 22. Slept at the 'Crown' – after Breakfast – Mr. Simpson accd. us round by his former farm at Stoneleigh, we were driven thro' both Parks, & close by the village to the Abbey, which is a mile away, forward thro' fine Oak timbered woods & Covers to Guy's Cliffe, the residence of Earl Percy, saw Gaveston's Tomb, &c – on our way back to Warwick, where we had Lunch with a famous Auctioneer, called Margetts – after which we attended the Show – saw the Horses paraded, the Danish 'Milker' – the American Shearing Machine – Shear 100 per day per man – called on Messrs. Sutton's, Thorncliffe's, &c – left 5.25 Shipton 8.25 after a *most* enjoyable visit, in extra fine weather!!!

June 26. Grace Catherine Calvertt & Fredc. Wm. Matthews 'asked Church' for the first time, at Fifield, & Leafield.

July 6. L.M. Wynne, working very hard, at electioneering, speaking 3 or 4 times & places, per day.

July 11. Shipton Auction – Finished Turnips in!!!
John Chas. drove Miss Ada Lowe, & Grace to Witney.
Abundance Strawberries, our very best crop.

July 13. Was driven at 7½ a.m., to vote for Mr. Wynne at Leafield School House – John Hadland, Wm. Sampson, Robt. Caldicott, Thos. Dore, George Steed, Chas. Pratley, Thos. Lock, & Jas. Pratley accd. me in a Waggon.[4]

Sir B. Samuelson	3640	North Oxon.
L.M. Wynne	3453	
	187	
Honble. Frank Parker	3688	South Oxon.
Sir Walter Philimore	3269	
	419	
Mr. Benson	4278	Mid Oxon.
Mr. Morrell	4167	
	111	

A.H. Smith Barry, 22 Majority in Sth. Huntingdonshire. I did not reach Huntingdon until 6.3 p.m.

July 19. Grace Catherine Calvertt, became Mrs. Fredc. Wm. Powell Matthews, Fifield, Chipg. Norton.

Isabel, Miss Ada Lowe, Miss Matthews, Fifield, and a Miss Matthews, Taynton: were the 4 Bridesmaids.

The following are some of the Wedding guests – Revd. Mr. & Mrs. Lee, Revd. York – Mr. Mann (was Best Man) Mr. Mrs. Burbidge, 6 Matthews – Mr. Yapp – Mr. Mrs. Bliss, Mrs. Gask, May & Frank, Mrs. Emerton, Mr. F. Smith – Miss Mace ńee Mann[5] – Mrs. Abbott and her Son Bernard &c – L.H. Barker.

Married 2 p.m. Leafield Church – by 4.50 train for Inns Court Hotel – London.

Wind & rain rather marred the Ceremony, otherwise, all passed off nicely.

July 23. Mr. Davenport rode over on his Bicycle.

Election Returns –

Conservatives	267	Gladstonians	274
Liberal Unionists	47	Anti Parnellites	72
	——	Parnellites	9
	314		——
			355

Orkney & Shetland M.P. to finish!!![6]

July 26. Driven with Isabel, & called for Mr. Matthews, Fifield, had Lunch & farmed round Icomb with Mr. S.M. Simpson.

Icomb Place, & village, are very retired & picturesque, I admire Guy's Tower, very nice scenery, from its top.

July 28. Wrote Crown for Remittance for the 2 years ending next Michs.

Mr. Abbott by 8.10 a.m. train for Town: he expects to leave Liverpool on 11th Augt. – in S.S. 'Asia', for Bombay.

July 29. Finished carrying 120a. Clover – in good Condition!!!

Dragged Ascott Common Fallows, and Turnip hoeing well in hand, 108a.

Aug 1. Saw a Barge with 2 horses – a steam Barge – and Steam Launch, all start loaded for Nuneham – Bank Holiday – a great gathering in Blenheim Park. . . .

Aug 2. On Saturday Fredc. George reached Home from 1001 Bathurst Street, Toronto, in the S.S.'Oregon'.

Aug 5. Mr. Hayley Gregson shewed John Chas., Robt. Wm., & myself over Langley Farm – House – Cottages, &c.

Mr. A.E. Abbott started his Luggage &c from Ascott to Liverpool, whence he will sail for Bombay on the 11th inst. – calling at Port Said, & Suez, *only*.

Aug 6. Entertained 46 Labourers, their Wives, Plough Boys, &c – sat down 6 p.m. – rose up 10.30 after a very enjoyable evening – The Bride, & Bride Groom joined us, 8 p.m.[7]

Aug 7. Walked to Church with Miss Edith Abbott.

Walked down to Mr. Price's – found him well, after spending 10 days at Torquay, & Ilfracombe.

Aug 9. Had Champagne 11 a.m. – Mr. Mrs. Abbott, Miss Abbott, & Miss Mason, eat Bride's Cake. . . .

Aug 17. Posted my final letter to Crown Office – and offer for Langley Farm!!!

John Charles drove Miss Abbott, Miss Alice Mason, Robt. Wm. and Isabel to a Bazaar, & Fete, in Burford Priory Grounds.

Aug 19. . . . Bride's visit to my 3rd Daughter, Grace, Mrs. Matthews – saw 2 of their 'self-binders', at work, cutting Oats. Saw the school children, dance around Revd. Yorke's Maypole, recite, & sing, &c. – after a jolly good Tea.

Aug 22. Accepted the Crown offer, for Langley Farm.

Had 2 hours with Mr. Price in the evening –

Mr. Mrs. Fredc. Matthews drove over, supped, after much Thunder & Lightning, returned to Fifield.

Aug 26. Bampton Fair – 3 Reapers finished cutting Big Brazil – *All* hands began tieing Oats, & Wheats – 85a cut. . . .

Aug 30. . . . Met Richd. Stovin, Mr. Brassey's Huntsman, at Mr. Price's, he gave a very good account of future Hunting Prospects, expects to be Cub-hunting in Cornbury Forest on 9th Sept. & 10th.

Sept 1. Drove to Icomb Place, and joined Dr. Jay, & shot with Mr. Simpson & his only Son, Hayward. We shot 26 Partridges 5 Quail & several Rabbits: much of the Corn standing, saw nice show Birds, & a Legion of Rabbits.

Sept 9. Saw Hounds in Forest for their 1st morning – too busy to stay, this makes 4 good, *long* Carrying days!!!

Sept 20. A Thundery rain in early morn, & frequent heavy showers during the day.

Enter upon the Langley Farm 'Shooting', this day – after 11 Guns – i.e. – 5, 4, & 2 had been over the land by Mr. Gregson's invitation, unknown to the Crown!!!

Yesterday received a brace of Grouse from Invernesshire, kindly sent, by Mr. and Mrs. Dawkins, of Wilcote.

Sept 23. Robt. Wm. drove me to Bradwell Downs, had lunch with Mr. F. Smith, & attended Stock Sale at next Farm – Fat Ewes 28s. to 21s.!!! Best pen 5 Fat Lambs 34s. – Beast equally cheap!! Horses & implements sold fairly well!!!

Sept 29. On Tuesday my Grandson Bernard, crushed his finger in the Mangle – Dr. Roberts, spelled it up – John Chas. drove to Maugersbury & bot. 2 Cart Horses: of Mr. Jeffrey £69.

Sept 30. John Chas. drove Robt. Wm. & myself, to Mr. Parsons-Guy's Sale, at Begbroke: Mr. Partridge a Devon-ite has taken his Lease, 4 yrs. to run, off his hands, and he has been accepted as future tenant, by the Duke of Marlborough.

Oct 7. Mr. Mrs. Price drove over here for Tea – regret to say he becomes very infirm.

Oct 10. Rode to Shipton auction & saw Mr. E.H. Keck about Langley Valuation, & selling my 12 Cows & 12 Bull Calves, & 2 Bulls, at Chipg. Norton Junction Sale on 17th inst.

Called on Mr. Pain, Finstock: who I fear will never recover from his very serious illness: which I very much regret.

Oct 13. The Crown ordered Mr. Gregson, to enter upon High Lodge Farm, on the Comrs. acct. Mr. Bliss not having yet closed the Bargain! . . .

Oct 18. John Chas. & Robt. Wm., 'carried Harvest' – late Oats in Cockshot-hill – their 3rd & *last* – there.

Keeper shot Leveret at Finish, started by Fredc. Geo: & 'Victor' the Pointer – picked up & brot. Home by Fredc. Geo.!!!

Robt. Wm. flitting to Langley.

Oct 25. Shot my 1st Hare & Partridge on Langley Farm, Robt. Wm. also shot a Hare – saw more Game than we expected!!! Robt. Wm. slept at Langley House, for the 1st time!!!

Mrs. Bellchamber came, as House-keeper.

Oct 28. Rain – *very* mild – rain!!!

Nov 6. Walked down to Mr. Price's – he has lent his Pony to Dr. Roberts for the Winter!! I am sorry he cannot drive.

Nov 9. The 8th Duke Marlboro' found dead in his Bed – his only Son, will be 21 years old next Sunday!!! . . .

Nov 19. Met Hounds Eynsham Hall, had Champagne &c – Mr. Brassey sent me Brace Pheasants – shot at Dytchley!!! Called at Mr. Pain's, & Mr. Adams – after the Hounds found at Wilcote, & *crossed* the *Evenlode*, for Fawler &c. . . .

Nov 20. At Church in a.m. saw Mr. White who told me Lord Churchill had bot. Rolleston Hall, Northamptonshire with about 1000a. land – his Lordship now wishes to sell his 6000 acres, around Cornbury House, & Park.

Nov 26. Mrs. Abbott left by 1.15 train for London, on her way to Bombay, a journey of over 6000 miles!!! she expects to start on Decr. 8th from Royal Albert Docks, in P & O steam ship 'Paramatta', call at Gibraltar, Malta, Brindisi, &c.

Mrs. Price 2s. Mr. Price 1s. lost at Bezique.

Nov 28. Met Hounds Heythrop – ran to Honble. Mrs. Hall's Barton. Called with Mr. Bliss, Davis, Lockwood, & Edwards at Heythrop House – Hot Mutton, & Soda & Whiskey – Mr. Brassey showed us round the House – Stables – &c.

Dec 7. Met Mrs. Abbott 10.25 a.m., on Paddington platform: we shopped – called 200 Sloane St. Mr. Yapp out – visited 'Venice' at Olympia: dined 53 Guildford St., Miss Pulford's, and I went to Alhambra – Don Juan – & Victorina, the strong woman, she carried 865 lbs. weight: & a 500 lb pony off the Stage!!!

Dec 8. Mrs. Abbott & myself – reached Albert Docks at noon, walked all over the S.S. 'Paramatta', she is bound for Gibraltar, Malta, Brindisi, Port Said, Aden, Colombo, New Zealand and Australia – Mrs. Abbott changes at Aden, for Bombay.

Miss Alice Mason, showed me back to 18 Beaumont St. Portland Place, Mr. Yapp, & we parted at Kensington Stan. – I left Paddington 4.45 and reached Ascott 7.20!!!

Dec 14. Met Hounds, Adlestrop – found in Evenlode Spinney, ran to Wolford, Barton, Brailes, Long Compton, nearly to Whichford, turned & ran to Walk Gorse, and killed on 'Cold Harbour', 1 hour & 55 min: the *Best* run of Mr. Brassey's 21 seasons. 10 to 12 in at the kill – *all* the Field lost at times, I came away from Long Compton, with Mr. Josh. Reade junr.

Dec 28. Mr. George Pain, Finstock: died ae 71 – a very nice genial man – very much regret his loss.

The Honble. Miss Dillon, fell on the icy street in Chipg. Norton, & fractured her leg.

Dec 29. Rode to Witney –

The Duchess (Hammersley) Marlboro', is inclined to buy Cornbury Estate, 4000 acres: she possesses £100,000 in Cash, and has £25,000 per year Income!!!

John Chas., Robt. Wm. & Isabel – joined a party of 45, at Fifield.

Dec 30. Viscount Dillon, altho' owner of 96,000 acres land in Ireland, & his Dytchley estate in Oxon – has only £1500 per year at the *present* time.

Dec 31. So ends in a keen frosty 8 days Rime, the wretched year 1892: for Agriculturists: for 'Produce Prices', the lowest on Record!!!

1891 & 1892, have proved the most ruinous of my Life?[8]

1893

Jan 4. Mr. Mrs. F.W. Matthews, Miss Simpson, joined John Chas., Rt. Wm., & Isabel; & were driven to a Fancy Dress Ball given in Witney Corn Exchange, by Mr. Bateman – Grace as Moonlight; Isabel, a Nun – Miss Simpson a Gipsy – two Huntsmen, & [a] Cricketer.

Jan 7. Four inches snow on the level, after 14 days *hard* frost – Arthur John Hazelton, late Heythrop Huntsman, died ae 47 – on Monday last.

Jan 9. Had Lunch, and Tea, & Cards, with Mr. Mrs. Price. John Baughan entered Tenant for Mr. Price's Farm, 16s. per acre, 354a.

Mr. John Chaundy, applying for High Lodge Farm.

Sold Fat Ewes 20, & Tegs 10, at 44s. *net*. . . .

Jan 11. Already had the ploughs stopped by hard Frost, for 19 days!!!

Walked all around Langley Farm – Mr. Gregson 8 Carts hauling Stone for new Walls – 6 Digging Stone.

Jan 12. Walked down to Shipton, & spent 2 hours with Mr. Price & he told me about quitting his Farm; and Mrs. Price went & applied for a House, belonging the Reade's.

Jan 23. Called upon Mr. Bliss, & met Hounds at Heythrop, ran 5 Foxes, & killed brace.

Ploughing again today, for the *1st time*, since Friday 23rd Decr.

Jan 27. Rode all round, and called upon Mr. Price, found [him] unwell, after driving to meet Hounds at Gawcombe – in the cold Fog.

Jan 31. Parliament met yesterday, Lord Brassey moved the address &c.

This has been a very seasonable winterly month, and suitable for mellowing the land for Produce – which is cheaper than ever.

Feb 7. Mr. Beauchamp & Cresswell – offered High Lodge, to Mr. Chaundy, called here – & further planned works, at Langley, with Mr. Groves, & Gregson – & instructed me to arrange with Mr. Keck & Davids, about final finish to valuations on High Lodge, & Langley Farms.

Feb 11. Met Hounds at Chadlington – found Spelsbury Warren and ran by Dytchley House, nearly to Wotton & lost – Captn. Paull fell over wire & lay senseless for 2 or 3 hours unobserved – when found, carried to Honble. Ponsonby's.

Feb 12. At Leafield Church.

Walked to Shipton Post Office, and spent evening [with] Mr. Price.

Regret to hear Evans Huckvale, & Chas. Sargent, are arranging for going abroad!!!

Feb 14. Rain, and Fog: John Chas. caught 52 Rabs: Priest Grove. Read Mr. W.E. Gladstone's No. 2 Home Rule Bill – he spoke 2¼ hours on introducing it, to the House – it was a marvellous performance for a man in his 84th year!!![1]

Feb 25. Mr. Simpson, Icomb: met Hounds Cornbury Park – Dined here, & stayed night – Mr. W. Abraham met him – Met Hounds High Lodge Farm, tried Boynal & Priest Grove blank – Blindwell, blank – found in Singe Wood, ran to Forest – Lord Brassey told me, he was much amused the other night – with Ld. Randolph's speech, and Mr. Gladstone's reply!!! in House Commons.[2]

Feb 27. . . . Repairs & rebuilding, & Building new Walls, going on rapidly, on Langley Farm.

March 10. Rode to Shipton & High Coggs – attended late Mr. Adams funeral at Hailey,[3] and called & saw Mr. W. Wilkins Ramsden.

Grace & her Husband, attd. Mr. Pain's Furniture Sale, held at White Hart Hotel, Charlbury.

March 13. Shipton Auction.

Joseph Reade Esqre. died 4.30 p.m. – ae 77.

Mrs. Pain, Finstock House, sent me 2 Whips, and her late Husband's Spurs.

Billiards – J. Roberts, v. W.J. Peall, £200 a side – This spot barred match at Egyptian Hall, Piccadilly: has been in progress a fortnight, 24,000 up – Peall to have *9000 start* – J. Roberts won by 1635, made an *extraordinary* break, of 737 during the match – the grandest match ever witnessed!!!

Mr. Ed. H. Cartwright gives a Doleful account of Linc. Farming!!!

March 15. Mr. Price 70 years old to-day: Dined with him.

Drilling Beans & Peas, the last 3 days.

Dry, very bleaching weather, for the land.

March 17. Rode from Icomb and called upon Mr. L. Barker, Stow: joined Hounds at noon – found in Lower Swell osiers ran & lost – tried Copse Hill Gorse – and found our run Fox, in a Cover nr. Eyford – very soon killed – tried the whole of Eyford & Upper Swell osiers, blank, found in Captn.

Thursby's little Cover, ran to Upper Ashes, & left them running to Oddington Ashes.

Mr. Lowther Barker joined us at Dinner 6 p.m. at Icomb Place – had some very good Whist.

March 22. Oxford won the Boat Race 18m : 47s the *fastest* on record – in the inter-university athletic sports an Oxford man, jumped 5ft. 11in. High, and another 23ft. ½in. Broad.

'Cloister' won the Grand National, by 40 lengths, in 9m : 42s!!! the *fastest* on record.

March 24. Mr. P.S. Yapp, told me his Son Peter, was engaged to Miss Alice Mason, 18 Beaumont St., Portland Place.

School Comtee. Meeting, Leafield.

March 25. The Driest, Sunniest March, I ever remember.

Langley House roof off – and on again – repaired, and re-slated – Highway new Wall finished. . . .

March 30. . . . Mrs. Abbott sent me an Indian Guinea, 250 years old!!! and a small Time Piece, given her by the Marabaune of Baroda, on her way out to India, in the 'Paramatta'.

March 31. Good Friday.

March has been *continually* Dry, with Rime Frosts, and Sunshine!!!

April 6. . . . At Witney Market – Best Red Wheat 22s. to 23s. – Seed Oats 18s. Barley 25s. to 27s. – Best Tegs 7½d. slow – Beef 6d. lb.!!

April 14. A *gigantic* strike imminent, begun already at Hull Docks, & likely to spread to London, Liverpool, Goole, Glasgow, Swansea, &c: the sooner the better, that Capital & Labour disputes are settled – men refusing to work with *Free* Labourers – as against *unionists.*

April 18. Heard Cuckoo – 1st time.

Cherry Trees in full bloom – very early.

April 19. Primrose Day.

Drove over, and attended Chipg. Norton, Board of Guardians, 1st meeting of the year: The Revd. T. Harris, in Chair.

Mr. Brassey advocated a Steam Roller for Highways.

Mr. A. Groves called & we chose Papers for Langley House walls.

Saw 1st Swallow!!! . . .

April 26. . . . I had 10 hours at Langley – Messrs. Groves placing new Terra Cotta Chimney Pots on Chimneys –

Messrs Leigh, fixing new Kitchen range – and Messrs. Ferriman, finished new Wall next 'Notgrove'.

April 30. Nine weeks to-day, since rain fell to any amount – a Drought prevails mostly over the whole of Europe –

Pine-apple, Bananas, & wine, with Mr. Price.

May 4. Mr. M. Savidge's Shorthorns, sold by Mr. John Thornton, 57 Cows averaged £20.16s.0d. – 18 Bulls averaged £21.18s.1d. Large Company, Lord Moreton, in the Chair.

Tea'd & spent the evening with Mr. Frank Craddock – Messrs. Wells, Hiorns, Wyatt, Corbett, & Hewer, there.

May 6. Sir Nigel Kingscote, & C.D. Beauchamp, came over the Wychd. Estate – Examined Langley House, Farm, & yards, repairs, &c.

Mr. Davenport called & I showed him over Langley House, on his way to Shipton Stan.

May 7. Brilliant sunny morning, with cold East Wind – not had a rainy morning, for *10 weeks*!!!

Strawberries, Hawthorn, Laburnums, Azaleas, been in full Bloom for days.

Prince George, the Duke of York: to marry Princess Mary of Teck: to the satisfaction of *all*!!!

May 10. Her Majesty opened the Royal Imperial Institute – one of the grandest sights ever beheld in London – her 3 Sons, & 2 of her 3 Daughters, were there.

Cutting down, & trimming the Trees, & Scrubbs, around the House at Langley, to the deep regret of the Sparrows!!!

Messrs. Parsloes, Painting, Papering, &c.

May 11. The Crown finally settled High Lodge, & Langley Valn. claims, by allowing £750 off the Fairspear, & Highlodge Rent, of the past 2 years.

May 16. A nice rain, during the night.

Mr. Yapp drove me to Fifield – shot 8 Rooks, then rain – £560 subscribed by the Heythrop Hunt, for a present to Mr. Brassey, for his already 21 years Mastership.

May 17. Digging up half the neglected Garden at Langley in front of House.

Heavy Thunder rain, at noon.

May 19. Frequent rains, day and night, since Monday afternoon, splendid growing weather once more – after being mostly Dry!!! from Feby. 26th ult.!!!

May 24. Spent the evening with Mr. Price, at Shipton Club Feast, Leafield Band – our last together!!! oh! dear, what constant changes, thro' Life –

Her Majesty 'Victoria', 74 years old!!!

Live in hopes that she may reign 60 years, & surpass her Grandfather, Geo. III!!!

May 26. Leafield Club – dined with them at the 'George': & Revd. Lee.

May 28. Spent the morning in Oxford 'New College Gardens' could sit & hear the Organ & Singing, in Queens – the weather magnificent.

May 29. Down the Isis to Iffley Lock, walked over the 'Isis' Inn.

Witnessed Somersetshire play cricket v. Oxford, in University Park – Hewett made 96 – score 317.

Attended the New Theatre – A Lass who loved a Sailor – A Bishop & Burglar, & Helen of Troy, were performed.

May 31. Mr. E.H. Keck, Appraiser, Bourton on the Water, died on Monday 4.30 a.m. – ae 73 to my great regret.

Attended Chipg. Norton, Board Guardians, Sanitary Comtee. &c. . . .

June 3. I retd. to Icomb with Mr. Simpson, farmed him round, had Gingerade at Icomb Tower – rain set in!!!

June 20. Mr. Gregson cleaning out the Moat around High Lodge – Robt. Wm. & John Chas., carried Clover 3 days: in good Condition. Mr. Fredc. W.P. Matthews, laid up of Quinsey.

Posted the Crown the Bal[an]ce of Rents, Tithes, &c – as between quitting High Lodge, & entry on Langley, Michs. last!!! £308.8.6d.

June 21. Never had so many Cherries & Strawberries: the Crops and Quality are magnificent.

Yesterday Her Majesty completed *56 years* reign!!!

Rode over, and Tea'd & farmed round with Mr. Sutton, Wilcote Manor.

John Chas. & Robt. Wm. drove over to a Garden Party, at Mr. Willm. Worley's, Witney.

June 22. Drove to Witney – left Mr. Abraham, Cherries & Strawberries. Saw Mr. Davenport about our Banking Affairs.

Heard Mr. F.W.P. Matthews, has Scarlet Fever. . . .

June 24. Sent Cherries & Strawberries, to Revd. Lee – Mr. Bliss, Price – never had such abundance of them.

July 6. . . . Tea'd at Revd. T.W. Lee's, with the School Children, a very enjoyable evening for all of them – Bazaar, Cricket, Football, &c. . . .

Horse-drawn tram, Walton Street and Kingston Road for Port Meadow, Oxford, 1890s. *(Oxfordshire County Libraries)*

Laying the foundation stone of the new town hall, Oxford, 1893. *(Oxfordshire County Libraries)*

Mr. Lucas, Mayor of Oxford, laid the foundation stone of the new Municipal Buildings.

The Duke of York, only Son of Prince of Wales, married Princess Mary of Teck – at St. James' Palace Chapel – the Queen att[ende]d – weather magnificent – the Grandest Wedding ever seen in London – the Wedding Breakfast at Buckingham Palace – hence to Liverpool St. Station – to York Cottage, Sandringham, for the Honey-moon.

The King & Queen Denmark – Czarewitz – Prince Henry Prussia – Prince Albert of Belgium – The Prince of Wurtemburg &c – were present: besides *all* our Royal Family.

July 10. Shipton Auction – very bad trade – owing to the *great scarcity* of Keep – from the continued Dry weather, since February!!!

Mrs. Ward, & her 2 Sons, came from Kensington.[4]

July 22. Had a day's navvying at Langley – levelled wall foundations, covered newly made stone road over the Paddock – clayed under the Water Trough &c – &c – cleared up all in readiness for Harvest.

July 25. Childrens' Fete, at Capps Lodge –

Mr. Price drove me to Mr. Brassey's Ram & Ewe Auction, Broadstone Hill – rams only averaged £8.12.10d. Shearling Ewes 58s. – Old Ewes 47s. – large Company: Lunch & Tea.

July 26. Began 2 Reapers Wheat cutting on Fairspear, 10a. Peas, & 8a. Peas, cut on Langley, & Fairspear.

July 31. Mrs. Ward & 2 Sons, retd. to West Kensington by 1.15 train, after spending 3 weeks here.

About 94a. Wheat, 18a. Peas, & 8 Oats, Reapered already!!!

Aug 1. The German Emperor in his Yacht 'Hohenzollern' reached Osborne on Saturday 3 p.m. – for a week's visit. . . .

What a difference in seasons!! finished cutting 108a. Wheat on Fairspear to-day – last year, only *began* wheats, on Augt. 25th.!!!

Aug 7. Began Wheat Carrying, 2 Ricks up in Big Field – & Peas.

Grand Temperance Fête, in Eynsham Park, and Foresters Fête in Blenheim Park.

Aug 11. . . . The *Weather* has so much to do with Cricket, fancy the famous 'Surrey Team' *all out* for 34 at Lords!!!

Aug 16. Hot scorching weather.

Isabel, returned from Newton Abbott, & Swindon.

Mrs. Matthews prematurely confined of a Daughter, at Lynmouth!!! after escaping a Coach accident – still born.

Aug 19. Returned from Icomb, called at Fifield, & bot. Grapes for Isabel – A gigantic Coal Strike in Wales, &c.

Aug 23. Finished carrying *all* at Fairspear, except 20a. Barley, and 10a. late Oats – only *began* last year on *this date*!!!

Aug 24. Mr. Matthews, & Mr. Jacob, only bid me 24s. per qr. – for John Chas. old wheat: 30 qrs.!!!

Mr. Matthews told me, Mr. John Baughan, would not be able to take Mr. Price's Farm!!!

Aug 26. Were driven 9 miles to East Ilsley Sheep Fair – large show Sheep – but few Buyers – slow trade, rather, scarcely any trade at all!!! about the worst I ever saw!!! Ilsley is 9 miles from Newbury: reached Ascott 4.15.

Aug 28. Florence Ferriman, confined of a Daughter at Langley, on 24th.

Water becoming very scarce!!! all around this locality, and well it may.

Aug 29. Mr. Robt. Abraham sent us 16 Peaches: saw his Self-binder at Work –

Mr. Fras. Gask, drove Mr. Walter Martin here – who told me the Linc. news: Mr. M. farms 1300a. at Wainfleet – some of the best arable land in England!!!

Aug 31. Mr. Richd. Eagle bot. Bull £16.

Messrs. John Maddox, Chas. Sargent, & Evans Huckvale: are all re-taking their Farms, at reduced rents.

Sept 1. Sent Mrs. F.W.P. Matthews a br. Birds, to Lynmouth, by *parcel* post 4½d.

Sept 9. By 8.15 train to Oxford – Best new Red Wheat 27s. – Barley 23s. to 33s. – Oats 18s. – a wretched Harvest!!! neither quantity nor quality.

Sept 11. The worst trade at Shipton I ever witnessed, especially for Store Stock.

Sept 13. Isabel accd. me to Mr. Doyne, the Occulist, 'Kilcrone', Woodstock Road, Oxford – called on Bride 'Dorrill', née Miss Elizh. Davis, Ascott.

Sept 16. Dry as ever, extraordinary weather, a 'Water Famine', prospect!!

Called on Mr. Price, & Davis. . . .

Sept 27. Robt. Wm. became a Juryman to-day.

Mr. Matthew Savidge's Sale of Sheep, Pigs, & Implements – the Short

Horns were sold in May.
Lunch & Tea, with Mr. Craddock, Lyneham.

Sept 29. Mr. Price's Stock & Implement Sale – prices satisfactory, as Times are – Ewes 37s. to 43s. – Ewe lambs 25s. to 30s.6d.
Leafield Harvest Thanksgiving.

Oct 3. Joined Messrs. Davis & Bliss, walked over High Lodge, I only shot Bird & Rabbit . . . Dined at Mr. Davis's – *Scotch* haunch Venison!!! Bliss, John Chas. & Fredc. – an enjoyable evening!!!

Oct 30. Accd. Mr. Price, with Mr. Matthews, & Lane to the grand opening Meet at Heythrop – 2 large Silver Wine Coolers, and a large Silver Salver presented by the Hunt, names of subscribers cut on the Silver Salver = 284.
Mr. Freeman-Mitford presented them – about 1000 persons at the 'Meet' – a Mr. Hone fell, and had concussion of the Brain – a man from Charlbury had his leg broken by the kick from a Horse.
Had nice running – killed a Fox.

Nov 1. Drive to Chipg. Norton Fair – attended Board Guardians, Assessment Comtee. Langley Tithe passed.
Corn & Sheep better trade – Sold Mr. Matthews 57 qrs. Wheat, and 30 qrs. Barley, 27s. per qr.!!!

Nov 3. About to plant 40 Yew trees, & 43 kinds of Fruit trees, at Langley – Farmed round with John Chas. & Robt. Wm.
Rode to Mr. Price's, Parsloes & Groves, and on to Messrs. Matthews, Fifield: Miss Lucy Robinson, Anderby, there – looked at the new House, just begun to put on Naunton Slates – heavy, but durable!!!
Hunting moderate, Foxes *plentiful*, around Stow.

Nov 4. Attended School Comtee. meeting at Revd. T.W. Lee's.
Tea'd & spent evening, with Mr. Mrs. Price – Mr. Cheatle called. Mr. Clement Pinnell fell off his Horse, and dislocated his Shoulder – Miss Reid, Shipton Court: fell over a wall, and broke Collar-bone.

Nov 11. Met Hounds, Wilcote: called Mr. Sutton's – Holly Grove, blank, found in G. Dawkins – ran thro' Forest to High Lodge & back to Finstock – found in Blindwell – no scent: tried Langley Groves, Priest Grove, & Boynal, blank: Home: Weather dry, & cold, and scentless, as March!!! an extraordinary 9 Dry months.!!!
All kinds of Agl. work, in a very forward state.
Mr. Lardner, Ascott: died ae 75, on 1st Novr. ult.

Nov 13. Mr. Mrs. Price, left Shipton by 8 a.m. train for Newbury, where they will reside in Future – I lose a very Good Neighbour!!!
Shipton Auction, Tea'd with Mr. Brookes.

Nov 20. . . . Mr. Howse died on Saturday . . . Leafield Blacksmith.

Nov 27. Quite a hard frost – called & found Mr. Mrs. Bliss in bed!!! met Hounds Heythrop – waited till 11.30 – killed in new Cover left them nr. Chalford Green 1.30 bad scent.

Nov 30. At Witney – Brisk trade for Sheep, & Beast, wretched for Corn!!! nice mild weather.

Dec 2. Rode Blenheim Palace 11.30 met Hounds, had Breakfast in Entrance Hall, where the Grand Ballers, had supped the previous night, and where all the Tenantry dine to-night!!! The young Duke looked remarkably well – as did his mother, the Marchioness of Blandford – sharp frost during the night, and froze all the morning: R. Stovin decided it was too hard for hunting, so at 12.30 we returned Home!
Dined with Mr. Davis, Ascott, in Evening.

Dec 3. Still, Dull, frosty morning. Wrote Mr. Price.
Mr. Addy leaves South Lawn, next Michaelmas.
Duchess 'Hammersley', intends to be married a 3rd time.

Dec 9. Met Hounds at Dean – called upon Mr. Mrs. Bliss, Chadlington, the Bell was tolling for Mr. Barton, the Father of the Heythrop Hunt. . . .

Dec 16. . . . Fine mild morning – abundance of Violets, & Birds whistling like April!!!
Hounds met Ascott, found in Langley Grove: ran to ground in Dix's drain Leafield – found 2nd in Widley, ran to Milton quarries, Taynton quarries, Tangley, Fifield, Bruern, Milton, & to a drain near Shipton Iron Church – very good day.
I attended Oxford Xmas Market – bot. Misses Hurdman, Cards.

Dec 27. Rode to Chipg. Norton – & sat 3 hours on Board Guardians, Sanitary & Highway Board, allowed £4 – for Langley Highway improvements. . . .

Dec 30. Forwarded the signed agreements, of Fairspear & Langley Farms, to 9 Whitehall Place S.W.
Met Hounds at Blenheim, & had my *first* day's hunt, with the Duke Marlboro' – large field – bad scent – Blk. frost, Rime frost, & occasional Fog – Earl Jersey, & Visct. Valentia there.

Dec 31. Trees look very pretty, laden with frozen Fog, with brilliant sunshine.

Mr. Mrs. F.W.P. Matthews, came & attd. Church, and Dined.

Of *all wretched* years, in Agriculture, this takes the *Cake*!!!

1894

Jan 4. . . . *Very cold* north east wind, & slight snow, every appearance of a *severe* storm: ground frozen *very hard*, and now a covering of snow – looks like keeping the Ploughs idle for some time!!!

Jan 5. Influenza – very prevalent at Shipton.

Jan 8. Shipton Auction – too cold & rough, for my going!!!
Thursday & Friday last, were the 2 *coldest* days during the Century!!!

Jan 11. John Chas. drove me to Witney, dined at Marlboro' – Sold some Blk. Oats 19s.6d. was bid £5.10s. for Mixed Clover Hay per ton, at Ascott Stan. – *Best Red* Wheats 26s. . . .

Jan 13. Mild rainy morning – & much Fog & drizzle, during the day – Dammed up the large Pond at Langley, & presume it will be *many* years, before the bottom of it, is seen again!!! I shall be very harmless, & quiet, in my Grave.

Jan 15. Met Hounds at Heythrop – very foggy – found & killed in New Cover, found 2nd nr. Kitegrove, lost amongst the Laurels nr. Great Tew House & Church – found 3rd in Worton Heath, ran a ring over the Country – chopped one in a Wheat field!! & lost our Fox in Worton – found again in Conegree, raced him 20 min – & killed – large field – retd. with Mr. Bliss by Chadlington Manor.

Jan 23. Had my *last* day with Dore, trimming Bushes, and cutting the Black Thorns, by side of Natteridge – on Langley – Trees & Bushes, now look uniform – Ponds cleaned out – Fruit Trees & Shrubs in Garden planted – Highway cleared – Big Barn repaired – Walls rebuilt – House & Buildgs., repaired – no extra expence required on Langley for years.

Jan 26. Two brace very fine Pheasants reached here from J.F. Mason Esqre. Eynsham Hall.
Had ½ an hour with Mr. Mrs. F.W.P. Matthews on my way to meet Hounds at Tangley – found in the Coombe – found in Taynton Quarries – very moderate scent, cold clean freezy air – left them at 1.30. . . .

Jan 28. Fine, cool, bright, sunny morning – At Church.
Wrote a long letter to Mr. Price.

The graves of Jane Waters Calvertt (left) and John Simpson Calvertt (right), Leafield Church. *(Colin Miller)*

Langley House, Langley Farm. *(Colin Miller)*

Mr. Lowther Barker, Stow, came and Dined, & told me he had sold his Establishment to a Brewer at Bath – and quits 24th Feby!!!

Feb 1. At Witney – Wool 22s. – Wheat 24s. to 25s. – Oats 19s. to 21s. – Barley 26s. to 30s. – Best 10 st. Tegs 5s. yesterday, at Oxford.

Walter Early, drowned himself – had 2 brothers in Asylum – they are one, of the two, Blanket Manufacturers of Witney.

Feb 3. By 1.15 to Oxford – John Chas. & Fredc. met Hounds, at Shipton Barrow, found 5 Foxes in 'Stockley Wood' & killed 3 of them.

Miss (red haired) Evelyn Brassey, became (to-day) Mrs. Leveson-Gower.

Feb 5. Mr. Mrs. Price, Donnington Villas, Newbury: like their new Home –

Mrs. Price walked me round Newbury, Shaw House, &c – Walked round by Sheen, Donnington Grove, Donnington Castle Ruins, Donnington Square, &c.

Mr. Mrs. Gibbs, late of Leafield, & 2 Daughters, reside on the Oxford Road.

Feb 7. Left Newbury 11.28 – reached Ascott 4.15 – after a very pleasant visit, with my old 18 years Neighbour!!!

Feb 12. Met Hounds, Heythrop: killed Fox in New Cover, found in Enstone valley & ran thro' New Cover, and past Mr. Kenny's, down the Park to Church, and thro' several of the Covers to Little Tew Lane, over the Country to Fullwell, turned back over Falkner's Farm & thro' Enstone, down Lidstone valley, & over the fields to Chalford Green, then nearly to Heythrop Park, turned round back, & lost nr. Chalford Green – 2 hours!!

Called at Mr. Blisses, Chadlington.

Feb 19. John Widdows, Standlake, dead, ae 50.

Mrs. Clinch, Witney, ae 67 died on Fri. 15 ult.

Mr. Willm. Worley, Witney, ae 65 died on 16 ult.

Feb 28. Met Hounds, Chipping Norton Junction, found in Starveall, & ran to Tunnel in Mr. Huckvale's hill Farm yard – found 2nd Fox in Gawcombe, ran to Wyck, & Little Rissington, & killed in Mr. Matthews' Pig Stye at Fifield.

My Horse, put his foot in a hole, in galloping over Fallows in Little Rissington, came down, and rolled over me, broke Collar Bone, injured Neck, Throat: & very much crushed Arms, Shoulders, &c.

Mr. Cheatle & Lane, set my Bone, or rather bandaged it up – had much pain for a week – dreaded going to Bed.

February has been a very fine mild month.

March 15. Not been able to write for 15 days.

Mr. W.E. Gladstone, resigned Premiership, to Earl Rosebery.[1] . . .

Mr. Wallis Thos. Wells, Withern, Alford, Linc. – started a Bull from Claythorpe & Aby Stan. G.N.R. – at 7.40 a.m. – he came by Lincoln & Worcester, & reached Shipton 8.19 p.m.

March 20. Mr. Cheatle came, & took off Bandage, & set my right arm at liberty after being fastened to my side for 3 weeks!!!

March 22. Was driven to Witney – Beef only 6d. lb. – Mutton 7d. – Wheats 23s.

Edmund Temple Godman, Banksfee, died 3 a.m. – Pneumonia – ae 49. He married the Honble. Susan Verney a Sister of Lord Willoughby de Broke, and has only *one* Son ae 7.

Buried at Longborough. . . .

March 28. Mr. Davis, Ascott, summoned John Hambidge, for shooting his 'Fancy' Pigeons, off Mr. R. Hambidge's Buildings.

April 6. My Collar Bone, hand, & Shoulder, are still very painful.

Sold Mr. Matthews 30 qrs. – dry sound wheat 24s.!! per qr.

42 years to-day since I commenced farming at Tothill Manor, nr. Alford, Linc: under Lord Willoughby de Broke!!!

5 men planting Potatoes, Peas, &c: in Garden – Dry!

April 8. Walked down to Mr. Davis's, Ascott: saw & heard the Cuckoo. Mr. Davis & Rowell, persuaded me to write to Mr. Bennett of Wroxton, nr. Banbury, to meet me at 'Blue Boar', Chipg. Norton next Wednesday.

April 11. Drove to Chipg. Norton, met Mr. Bennett, a famous local Bone Setter – found my right Shoulder out – had been 6 weeks – knew something very sadly wrong, but thought I was crushed only – The pain was extra *Grand*, when Mr. Rowell & Bennett pulled my Shoulder in, bandaged – & ordered Rum & Neatsfoot Oil mixed; to be applied *daily* to *Shoulder, Arm, & hand*, for a week, and meet him again.

April 14. Potatoes, & Garden seeds all in – the Plums & Pears in full bloom – early Potatoes – ready to hill – Lettuce & Rhubarb ready for use in the open. Peas, Onions, Carrots, all up – all young fruit trees, planted in the Autumn, alive and mostly in Bloom!!! altogether, never saw everything in the Garden so early for April 14. . . .

Nice showers are daily & nightly falling, every *present* prospect of an *Abundant* Season.

April 18. Attended School – and Franklin fund-meeting, at Revd. Lee's – Examined late Mr. Lardner's Cart-horse at Ascott, John Chas. bot. a pair,

£63 – I was driven, and met Mr. Bennett from Wroxton, at the Blue Boar: my shoulder going on fairly well – splendid Thunder Rain!!! . . .

April 23. Called upon Mr. White, Waterman's Lodge: who told me Cornbury Estate, would be sold this year – or pass into the Mortgages hands!!!

April 27. Walked round by Addison's Water Meadow walk at Magdalen College, saw the Jackdaws, peck hairs off the Deer, in the 'Grove', for their nests!!! Boating, all the rage. Soldiers passing thro' – Municipal Buildings already 12 to 15 feet high – Oxford is now looking most lively & pretty – all in full term!!! was very amused with the performance of 'County Councillor'.

April 28. Met Mr. Bennett at Golden Cross – Shoulder & Collar Bone, going on to his satisfaction – left 3.35.

Isabel retd. from Icomb – Miss Simpson rode over.

Vegetation, said to be 10 days earlier, then the 'average'.

May 1. Fine keen bright morning.

At 5 a.m. the Magadalen College Choir sang their annual hymn on top of Magdalen Tower, whence a beautiful view, of Oxford City, can be admired.

May 11. Wrote Mr. Price.

Recd. 1st letter from my 7 years old Grandson Bernard Abbott, Greenfield House, Waltham Cross, Herts.[2]

Fine rain in afternoon.

May 12. Fredc. Geo: retd. from Coventry for a week – Rudge Co.[3]

May 13. Dined with Mr. Mrs. Davis at Ascott –

The famous Blisses Chipping Norton Establishment, sold to the Metropolitan & South Wales Banking Co.!!!

May 18. Leafield Club Feast – dined at the Fox: three Clubs, 3 Bands, at 'Potters Arms', 'George', & 'Fox', Hotels!!!

A new Lightning Conductor being placed from Ground, to top of Spire; direct height 144 feet – done by Mr. Blackburn of Nottingham.

May 21. Sharp frost – Ice an 8th of an inch thick – cut down the tops of early Potatoes – I fear injured a deal of Fruit and tender Vegetation.

May 24. The Queen's 75th Birthday, – at Balmoral.

John Chas. drove me to Mr. Simpson's Icomb: conversed on future matrimonial affairs!!![4] . . .

May 25. Saw the new Copper Cockerel placed on top of Leafield Spire. . . .

May 30. Splendid showers yesterday, which all Vegetation required: the Frosts of 20th & 21st ult. – have injured the early Potatoes, and cut off an *enormous* lot of Fruit of *all* kinds: most *annoying*: talk of *Fruit growing* in this *wretched, uncertain* Climate – is a Farce!!!

May 31. Drove to Witney, and consulted Mr. Westell, about Highway Trespass, with Horses, & Beast.

Best Red Wheat 23s.6d.: Beef & Mutton 8s. st.!!

May has been cold & frosty – quite checked our extra early Promises, of an extra early season!!!

Turnips & Mangolds up at Langley – 18a. in – Mangolds up at Fairspear – 14a. Mangolds, & Turnips drilled – 20a. ready. . . .

June 4. A *fine heavy* Thunder rain, from 5 to 8 a.m. – then showers, suit vegetation, beautifully!! – expect to see things grow now in *a week* – as they did all May!!! . . .

June 5. Mr. Willm. Abraham, came & farmed round.

Mr. Nathaniel Stilgoe, when driving to Banbury on Thursday, in meeting a Bicycle, his horse shied, the Backband broke, trap dropped down, & he fell into the Horse's legs, and was so very much injured that he died in an hour: he married Mr. M. Savidge's only daughter, of Sarsden Lodge.

'Ladas' won the Derby.

June 11. Shipton Auction – Mr. Matthews drove me over Langley & Fairspear & spent evening.

Great disturbance between John Mace, Fulwell Farm: & Mr. Garne, Burford: Mr. Garne makes his Nephew 'Bankrupt', thro' some *Lawyer Thief*, at Oxford; for a paltry debt of £17!!!

June 15. Mr. Brooks bot. 2 fat Heifers £36.15s. net.

Drove & farmed round Fifield, with Mr. Matthews & looked over the New House, which is nearly completed.

Nice rain at night.

June 19. By 9.7 train to Reading for Ascot –
Called on Comins – saw Coaches drive in – met the Royal Procession; Prince, Princess & daughters of Wales, Duke and Duchess Saxe Coburg Gotha, Duke Cambridge, &c – I had a good view of all the Proceedings from Ld. Ribblesdale's Stand – Weather and Company – Grand.

June 22. Thousands of acres of Land are for Sale, all over the County: Cornbury Estate, Black-Bourton & Ardley, Little Burford, Little Tew – 'Fulwell Farm' in Wootton – Swerford – Fairford, &c.

June 23. . . . At 9.55 p.m. The Duchess of York [was delivered] of a Son, and new King of England, the Queen has now three representatives for the Throne!!![5] . . .

June 28. Lunch at Witney, with Mr. Mrs. Davenport.

Julian's Circus paraded the Town, did not envy their dusty ride up & down the Streets!!!

June 30. Rode over Mr. Jocelin's Farms, at North Leigh, and Tea'd with Mrs. Gask – Mr. Yapp & Mrs. Johnson, called, on their way to Handborough Stan. – had a very *Fine*, dry week!!!

July 2. The Prince of Wales opened the new Tower Bridge over the Thames – been 9 years constructing – cost £1,200,000. . . .

July 7. . . . Mr. Richd. Eagle – came – but did not buy the 15 steers in 'Well Close'!!

Mr. John Joslin, Curbridge, nr. Witney – bot. 2 farms at Ardley nr. Bicester, 522a. 3r. 13p., and A331. 1. 30, with 2 Houses, Buildings, & 6 Cottages: for £4850 – out-lying land of Duke Marlboro', and the cheapest I ever knew sold, altho' a 2s.2d. per acre Tithe!!![6]

July 31. Grace came & accd. Isabel, to the Children's Temperance Party, held at Potters Hill.

I shot 11 Woodpigeons, Robt. Wm. 9: attacking his Wheats. The past has been a very satisfactory month: for all kinds of Vegetation: there is abundance of everything all over the Country – if – the next *2 months weather* proves fairly satisfactory, for *ripening, & securing* it!!! Game will be plentiful.

Mr. Peter Yapp is a great Benefactor, to the People of Leafield.

The Princess of Wales, and Daughters, leave England to-morrow, to visit the Czar at St. Petersburg!!! and attend wedding of Grand Duchess Xenia.

Aug 4. Beautifully fine, for ripening the Harvest.

John Chas. drove Miss Simpson to Witney: also arranged repairs, painting, & Papering &c – certain rooms in the House.

Aug 6. The Great August Bank Holiday – quite a Harvest for the Railways!!! My *first* night at Langley.

Aug 11. The German Emperor, visiting his august Grandma' at Osborne – Yachting – Reviewing – & entertainments, fill up his *very busy* time!!!

Nearly supplied Langley House, from Fairspear; which is being Papered, Painted, &c: for John Charles. . . .

Aug 16. Began reapering Wheats, on both Farms; in *Earnest*!!!
Mr. Chas. Hobbs, Bampton: gone to a Reformatory for 3 months!!!

Aug 19. Mr. Mrs. F.W.P. Matthews, drove here to Dine, our first Visitors, at Langley!!!

Aug 23. Robt. Willm. to Witney. Rain – rain.
50 acres Wheat, and 15 acres Peas, cut on Langley, and 50 acres Wheat, 28 acres Barley, & 10 Peas, cut on Fairspear.

Aug 24. Sir Nigel Kingscote, & C.D. Beauchamp: spent the day on our Wychwood Estate – Kitchen chimney top taken down & re-built at Fairspear – Iron Pipe from Langley House to Water Tank or cistern – Pumps & spouts repaired, at Langley Cottages.

Aug 31. Very busy cutting, & began Wheat carrying, on *both* Farms.

Sept 1. Rode 'Cymro' to Cornbury, and inspected the 6 days Sale of L.M. Wynne's Furniture, &c – his Lease expires 25 March next.
Cornbury House, Gardens, Park, & Forest, look charming: and *all* for sale these next 6 months.
Shot Grace, brace Partridges.
The two pairs of Stone Initials on each side front door Porch, are upwards of 700 years old!!! . . . H for Henry 2nd and E for his Queen Eleanor, of Guienne!!![7] Henry was Great Grandson of Willm. the Conqueror, & son of Matilda daughter of Henry 1st who was son of Willm. the Conqueror – Henry II began to reign 1154 and died in 1189 – it has always been customary for our reigning monarchs to place their initials, in Stone, on all Crown Land *Houses, Buildings* &c, erected during their reign: part of the *present* Langley House was built by Henry II – 2 pairs of Stone initials are also placed over a stone arched window, up the *Old* Staircase – they are great curiosities and show the antiquity of the House, an appanage to Wychwood *ancient* Forest!!!

Sept 5. Drove to pretty Icomb, shot an hour with Mr. Simpson & Son. Hail & rain – called for Isabel at Fifield on returning.
Weather very hindering for Harvest.

Sept 14. Rode to Chipg. Norton, with Langley voters' list –
W.G. Grace, the Champion Cricketer, has made 100, and upwards: 98 times: up to this date – in *first* class Cricket!!!

Sept 15. Had a grand week, for cutting & carrying Harvest – only 23 acres left to Tie & Carry out at Langley: & 50a. to cut, & 100 acres to Carry, on Fairspear!!! . . .

The initials of Henry II and Eleanor of Aquitaine, above the porch at Langley House *(Colin Miller)*

Queen Victoria 1879, on the farm buildings *(Colin Miller)*

Queen Victoria (V.R.) and Prince Albert (A.P.) 1858, on the farm buildings adjacent to the house *(Colin Miller)*

Sept 16. Delightful morning – At Church.

John Chas. came and Dined with us at Langley for the last time, as a Bachelor!

Sept 19. At 2 p.m.: Icomb Church, John Charles married Mary Hayward Simpson – Dean Leigh & Revd. Marston off[iciate]d – dry pleasant day; all passed off most satisfactory!!!

Sept 20. Sold King Clover Rick £3 per Ton.

At Witney Mkt. Wheats 16s. to 18s. per qr.!! Barley 22s. to 24s. – Oats 13s.6d. to 15s. – ruinous!!

Sept 22. Very busy trimming up at Fairspear, for Bride & Groom, Kitchen chimney repaired – Painting, Papering, &c – nearly done.

Nice rain at night, after a Dry fortnight!!!

Sept 24. By 1.5 to Oxford – sent 12 Journals amongst my Acquaintances – describing John Chas. wedding.

Witnessed 'Freedom' performed. Bot. a Gong 15s.

Sept 25. A very fine rain.

Municipal Buildings, Engine House, and new Corn Exchange, all building at Oxford.

Sept 30. Fine breezy morning – Barometer, extra high!!!

Walked to Leafield Church – saw Mrs. Peter Yapp, née Miss Alice Mason: and Miss Beatrice Gilbert, there.

Much of this month been very hindering for Harvest-work – 100a. Beans & Oats, to carry!!! . . .

Oct 4. Mrs. Simpson came from Icomb, with Miss Ellis: to Fairspear; to unpack a variety of Household Goods, on behalf of her Daughter – Mrs. Calvertt.

Oct 5. Finished carrying Harvest – I am delighted to write!!! Sold some Wheat 18s. – Barley 23s.6d. & Oats 12s.6d. net!!!? & Mixed Clover Hay 60s. per Ton: to Mr. O. King.

Oct 8. John Chas. retd. Home to Fairspear with his Bride after spending 19 days in London, Exeter, Penzance, Ilfracombe, and Bath.

Beautiful growing Autumn weather.

Oct 11. Drove to Witney – the *worst* corn *prices*, I ever saw in my Life – Wheats 16s. to 19s. Barleys 18s. to 26s. Oats 40 lbs. per Bushel 13s. & 14s. Peas 22s. to 24s. Vetches 4s. Bushel. Ruinous!!

Oct 14. Fine, keen, breezy North Wind.

John Chas., his Bride, & Miss Ellis, walked over here and tea'd, before attending Shipton Church with Isabel.

Oct 15. Willm. Goodall has Hunted the Pytchley Hounds 20 Seasons, & never had a day's illness, during the whole of that period!!! a Silver Salver & £1282 – has recently been presented [to] him, which is by far the largest amount of Money I ever heard of being subscribed for a Huntsman's post Services!!!

The Steam Ship 'Lucania', averages 21½ knots per hour, during her last 9 voyages to, & from New York: a pace, truly marvellous!!!

Oct 20. Left Shipton 8.40 via Honeybourne & Leamington to Coventry 1.45 . . . returning by Banbury & Oxford.

Fredc. Geo. lodges at 1 Hill Cross – he showed me the 'Rudge' & Ariel Cycle Shops – there are 22 Bicycle Man[ufac]tories in Coventry – saw 'Peeping Tom' &c.

We witnessed Mr. Irving Junr. & Co. perform 'Bunch of Violets' – High Speculation, & Robbery throughout.

Oct 21. Stayed night at Baker's Hotel, 24 Fleet St., Coventry.

After Breakfast walked past the Grammar School to end of fine Avenue of Trees, 3 miles long, out for Rugby, 10 miles hence – fine scenery – good land, mostly grass: walked into, & heard part of the Services, in St. Michael's, Trinity, & St. John's.

Coventry is 9½ miles from Leamington – 10 Rugby – and 18 Birmingham.

Admired the 3 'Coventry Spires': left 2 p.m. – reached Oxford 4.5 and Ascott 6.11 – Spent an hour with Mr. Davis – called Fairspear – & reached Langley 8.10.

Oct 25. Wheat & Oats, scarcely saleable at Witney – average runs of Barley 23s. – a wretched ruinous trade!!!

Oct 26. Dined with Bride & Groom, at Fairspear – a rather heavy rain – at last!!!

Oct 31. The Japanese, will lick the Chinese, into a cocked Hat.[8]

Much of the past month was dry – fogs – but no amount of rain, till last 10 days.

Dores, the Mason, resetting the splendid *old* stone Water Cistern, at Spring, near 'Large Pond'.

Nov 1. Drove to Witney – Wheats 14s.6d. to 17s. – Oats 11s.6d. to 13s. – Barleys 14s. to 25s. – Wool 21s.

Found plenty, and killed lease Foxes, yesterday, at Bradwell!! Wet – very

– & mild.

The Czar of Russia died at Livadia in the 50th year of his age – Alexander 3rd – succeeded by his Son Nicholas 2nd, who is in his 27th year.

Nov 2. By 1.5 train to Oxford & Newbury – found my Old Neighbour Mr. Price – very shaky.

Nov 5. Bot. Mrs. Simpson, Icomb Place – a Gong 17s. – at Messrs. Gills – High St. No. 5, Oxford.

Nov 7. Isabel journeyed yesterday to Leicester to visit her Sister, Mrs. Ward.

Nov 10. Mr. Wells, Charlbury: gave me £100 for 20 Tegs, 20 Ewes, & 7 Cull lambs –

Rode around High Lodge Farm – The Crown are now planting 25a. of Ascott Common, with Larch, Firs, Oaks, & Sycamore.

Mr. Mrs. Simpson visiting their daughter at her new Home – Fairspear.

Nov 14. Left Fifield 10.30 – the flood water very high at Bleddington Bridge – re-crossed it, with a Surgeon from Chipg. Norton who had railled to Junction to see a Patient in Bleddington – attended Board Guardians – Sanitary Comtee. & Highway Board – paid Langley Contributions, & 1s.6d. rate signed, by Revd. T. Harris, & J. Hampden Waller.

Water too high for crossing at Ascott – & found it very high at Shipton – rained nearly all day as well as night: not had so much rain in 48 hours, for some years.

Nov 15. Robt. Wm. drove me to Witney.

At 8 a.m. the Flood Water ran thro' Mr. Westell's House, at Witney Bridge.

Had 1873 Vintage Port, given by Mr. Bagnall to myself, & Mr. Matthews!!! . . .

Nov 18. A thorough Novr. morning, a wet, thick Fog, can only see about 150 to 200 yards – Barometer high!!

Nov 24. Labourers, their wives, & others, altogether about 50, were entertained to a 'Wedding' Supper, in Langley Barn, from 6 to 10 p.m.

Beautiful weather, after the Floods – no wind.

Finished stoning, & clearing out, the Langley Highway.

Nov 28. *Nine* months to-day, since my Shoulder was dislocated, & Collar Bone fractured – quite sound & well now. Went hunting again to-day for the *first* time since, met Hounds at 'Junction', found in Bowl Wood, lost in Icomb – found in Gawcombe.

Admired Mr. Samuda's House, Buildings, & Improvements at Bruern – found brace of Mr. Brassey's Phts. for me at Ascott Stan.

Called at Mr. F.W.P. Matthews, Fifield: and Mr. J. Davis's, Ascott Grange.

Fine, mild, still, Dull Day – scent varied.

Nov 30. Planted 12 Yews – an Apricot, & *Minster* Plum tree.

Inspected High Lodge Buildings – & the 25 acres planted with Trees, in Ascott Common – & Water improvements. Covered the Mangolds.

This month has been famous for *two* extra heavy rains, & *fine*, mild weather!!! . . .

Dec 2. 'Modern Socialism', is much of it, infamous, cowardly, & immoral, and to accept the Doctrines often advanced under its name, would be Politically a blunder: Commercially, a calamity, Socially a crime, & Morally, a sin: you may also define Socialism, as the Thoughts, Projects, & Schemes, which will advance & improve, the social condition of the People generally!!!

Dec 3. The Water Supply at Langley & Potters Hill, is now in *thorough* working order – which has not been the case for years!!!

The large Tank in Home field, behind the House, measures 19 feet long, by 6 feet wide – the small Cistern, 10 feet by 2, only.

Dec 4. To-day, Parish meetings are convened all over England, to put in force, the new Parochial Act.[9]

Henry & Chas. Pratley, & Geo. Steed, composed our meeting, they came to the conclusion we could not improve upon the Old System!!!

Dec 8. Met Hounds, Wilcote: found & ran 3 Foxes, bad scent, lost them all. Lunch with Mr. Mrs. Sutton.

Mrs. Dawkins ill in London.

Mr. Sutton become a Councillor, so is his Groom!!!

Dec 10. Attended Shipton Auction, only 375 sheep, & 21 Fat Beast – and a very slow trade.

At Oxford you can buy 2 loaves, 2 Spanish Onions, and 2 lbs. of Mutton, for 1s.!!!

Dec 13. Witney Xmas Fat-Stock Market; *Bullock* Beef, down to 3s.4d. per st. – No.7 £17 – judged at 100 stones!!!

Dined at Mr. Davis's, Ascott Grange, met Stovin & Drake – Hunt subscribe, £1400 per year.

Dec 21. Miss Edith Abbott, & Bernard, from Waltham Cross, to spend Xmas, at Langley –

Mist, Fog, & drizzle.

Dec 25. Xmas Day – Bernard Abbott, walked with me, to Leafield Church – Miss Edith Abbott, Fredc. & Isabel, also attended.
Still, mild day – wonderfully so.

Dec 31. Lord Randolph Churchill, *seriously* ill, from Paralysis!![10]
Robt. Wm. & Isabel, dined the Old Year out, at the newly built House, Fifield: belonging Mr. Matthews –
Fredc. returned to Coventry, reluctantly.
Been a further ruinous year, for Agriculture – Wheat averages 3rd week in Oct. 17s.6d. per qr.: the *lowest* on record!!!
Bullock beef at Xmas Auctions 5d. to 6d. lb. – Wheat *now* 19s. – Oats 13s. to 14s. – Barley 16s. to 25s.

1895

Jan 5. Lord Randolph Churchill, lies in a very precarious state, from Paralysis!!!

Had Frost, with ¼ to ½ inch snow, since Sunday morning.

My 3rd Daughter, Mrs. F.W.P. Matthews, Fifield: confined of a Son, his Grand-father Matthews's 54th Birthday!!

Mr. Harrold Matthews, wife & 6 children, left Taynton for Southampton *this* morning: for 3 months, then probably will go abroad.

Jan 6. John Chas. & his wife, walked over to Fifield, to see Grace and her Son & heir, Donald Marmaduke.

Jan 9. By 8.30 train, from Shipton to Chipg. Norton – attended Board of Guardians – made Declaration as District Councillor, with 30 to 40 others – Mr. John Addy in Chair – in Revd. T. Harris's absence.

Jan 16. . . . Drove to Fifield, and shook hands with my Grandson, now 11 days old – looks healthy – & like living in Old England for a long time. Mrs. Gask, & Miss Sotham there – gave Mr. Groves senr., Miss Dangerfield, and John Franklin a ride – on my drive to Ascott.

Jan 24. At 5.45 a.m. Lord Randolph Churchill died ae 45 – to the great regret of a multitude of Relatives, Friends, and political, & other, Acquaintances!!!

Jan 25. Rode to Fifield – saw Grandson – Lunch at Manor House, met Hounds at the 'Merry Mouth' – Large field, 5 Brasseys &c – found in Taynton quarries, I left them running towards Barrington Bushes, in a driving snow.

Col. Long: C: had 1175 majority at Evesham.

Jan 26. There was so much life in Ld. Randolph Churchill, he was so prompt, so ready, so active, & so bright: he conveyed the impression of so much pent-up energy, that it is difficult to associate his memory with the idea of eternal repose – his death is a cruel irony of Fate – at 45 he has had to bid farewell to all that he loved, to the world & its pleasures, to the noblest joys of public life, to faithful Friendships, to the tender companion whose devotion must have eased his long, and painful struggle with Death – in him England loses one of the most dashing and sincere, one of the boldest, & wittiest, of her Sons – He had many adversaries, but not one Enemy!!!

Mr. Winston, and Mr. John Churchill, are 'all' his Family – the elder, heir presumptive, to the Dukedom of Marlborough.

Jan 28. By 8.10 train to Handborough – walked & inspected Blaydon Church – Lord Randolph's open grave, on north side of the Church yard: went on to Blenheim Palace – then to Woodstock Church – at 11.45 a special train arrived from London, with Corpse, & Mourners, they proceeded to St. Mary's Church, Woodstock; then proceeded to the grave in Blaydon Church, where Ld. Randolph Churchill was laid at rest – after his short busy life.

The Bishop of Oxford officiated – Lady Randolph, & Lady Curzon, were the only two ladies of the Family there – her 2 only Sons – the Duke – and a host of others were there.

There are six Sisters, all strong, fine women – Duchess Roxburghe, Lady de Ramsay, Lady Wimborne, Lady Curzon, Lady Tweedmouth, and Mrs. Wilson!!! & their Mother, the Dowager Duchess of Marlborough!!

Dined with Mr. Davis, Ascott, on my return.

Jan 31. Robt. Willm. drove me to Witney – sold Mr. Munsey 50 qrs. Wheat 19s.6d.: and 100 qrs. Oats, 40 lbs. per Bushel 13s.6d. – Prices for Corn are ruinous – altho' we are only 70 to 80 miles off London – the *Best* Market in the World!!! but such is the Fact – the Carriage for a Ton of Straw, by Rail, from Shipton or Ascott, to London, is 16s.8d. – robbery!!!

January has been a very winterly month all through – Floods, heavy Snows, & severe Frosts, in many parts of England – especially severe abroad, even in Rome, and the Riviera.

Feb 4. Drove to Fifield yesterday – my Grandson Matthews was christened Donald Marmaduke – Miss Sotham gave him a Silver Mug, given her when christened 84 *years* ago – the mug is 120 years old. . . .

Feb 6. Great distress thro' scarcity of work – several able-bodied men applied for [Poor] Relief at Chipg. Norton.

Feb 7. Very keen frost – 2 inches snow on the level – Blue sky, with sunshine.

At Witney – wretched trade!!!

The 'Crathie', 450 Tons!! ran into the 'Elbe', 5,400 Tons!! off Lowestoft, & she sank in 20 min – !!! 334 were drowned.[1]

Capt. Leigh accidentally shot Col. Byng in the eye, at Cornbury: Dr. Doyne, & Higgins, & Maisey: had to take out the eye!!!

Feb 9. Rode to Mr. Wilkins, Finstock: Mr. R. Butler's, Charlbury: called on Mr. W. Wells, laid up of Pneumonia – also on Mr. Young, Ascott – Rabbits dying in Forest for want of Food, had nearly *6 weeks* frost – began on Dec. 30th ult.!!!

Feb 12. Mr. Joseph Reade, called: he presents himself as Councillor for County – opposing Mr. Saml. Groves, if he will fight!!!

Bright sunshine – ground frozen, with 2 inches Snow *on top* – Thrashing, or Wood cutting: are now the *only work*, Labourers can do – the Manure, & other hauling is all done!!!

Feb 15. . . . Jas. Francis Mason, of Eynsham & Freeland, married Lady Evelyn Lindsay, to the great satisfaction of *all* his Relatives, Friends, & even Acquaintances!!!

Three weeks to-day since Hounds met at 'Merry Mouth' – Frost as severe as ever, 1½ inches snow.

Feb 20. My Grandson, Donald Marmaduke Matthews, was driven here, for the *first time*, in his life!!!

I drove, & attended Chipg. Norton Board Guardians, about 30 there – spent an hour with Mr. Bennett, the Bone setter, Wroxton.

Mr. Brassey's County Ball.

Mrs. Jenkins, née Ellen Reade, became Mrs. John Arthur Rigge, Henley on Thames on Wednesday 20 this day!!!

Feb 28. Drove to Witney – sold dry, Red Wheat, 19s.6d. to Messrs. Holtom's – No ploughing or Hunting, done, during this Month!!!

Received 13s. per qr. for 60 qrs. Oats: Mr. Munsey; 19s. for 40 qrs. Wheat, & 21s.6d. for 25 qrs. Peas – Messrs. Holtom's; and 13s. qr. for 30 qrs. Oats, Mr. H.O. King: I name these, as being the *most ruinous* prices, on record!!!

This has been a very hindering, do-nothing month, for Horses, & men; plenty of work, but cannot do it.

March 1. Only about 30 Ewes to lamb: began last 4 days in Jany: & finish early in March: many years of my life began 1st March in Linc.

March 2. Land ploughs fairly well, where broken, or dry mold – for the *first time* since the *24th* Jany!!! Winter Beans dead – all Vegetables 'Brown', & dry, as gray paper!!! . . .

March 5. Forty four, half women & girls, are the *present* population of Langley: viz: 'Calvertt, Son, Daughter, & two maid servants, Chas. Ferriman, Wife & Daughter – Longshaw, Wife & Son – Stephen Moreton, Wife Son & 2 Daughters – Henry Ferriman & Wife – Steptoe, Wife two Sons & Daughter – Steed, Wife, three Sons & two Daughters – Dores & Wife – Hadlow, Wife, & Son – Jacques, Wife, & Daughter – Sampson, Wife, two Sons, & two Daughters.' – 44

Only three contested Elections, for County Councillors in Oxfordshire!!! J. Reade, & J.G. Abraham, unopposed.

March 9. L.M. Wynne failed to the amount of £250,000 –
Sir Bernard Samuelson, M.P. will not seek re-election –
Wynne has resigned, so North Oxon – is at liberty for a stranger – !!!

March 11. Rode to Burford & saw Mr. East – about Water supply.
Shipton Auction – heard Mr. Brassey has resolved to offer himself as M.P. for North Oxon – possibly unopposed!!!

March 15. Fine *fine* mild morning, *no* Frost, not had so mild an air, since *December 29th*!!! Wrote Mr. Price.
John Treadwell died ae 73 – Huntsman 19 years to the Old Berkshire Hounds!!! 1863 to 1882.

March 18. This makes the first 4 fine dry sunny days of *this* year!!!
Met Hounds at Chipg. Norton – found Fox below the Town nr. the Common – lost nr. Kennels: found in Daylesford, ran thro' Adlestrop 'Neats' – the Hill – to Compton, & lost at the Alcombes: tried other Covers till 3 p.m. – then started for 'Evenlode Mains', I came away – large 'field', 27 ladies out – reached Langley 5 p.m. – Mr. Mrs. Brassey, 4 Daughters, Lord & Lady Brassey, Leonard & his sister, Mrs. C. Leveson-Gower – Lord North-Leigh – Thursby – Freer – Rawlinson – Reade – Reids – Rice – Baker – Samuda – Bliss – Capt. Davis – Edwards – Neild – Schofield – &c – met Hounds at C. Norton.
Lord Brassey told me [he] should start for Victoria, Australia: in July, & take his Yacht 'Sunbeam', along with him, & should stay as Governor for 5 years.

March 21. At Witney – hired Bertha Dix – £14.[2]
'Spring Quarter Day' – nicely mild & still – a sign of a Fine Genial Spring!!!

March 30. Drove to Witney, & saw Mr. Davenport.
Showery, hail squalls, hindering weather.
Been very March, many weathers, throughout.

April 3. Met Hounds Bradwell – very nice running killed brace Foxes. Mr. Kingscote told me his brother Sir Nigel – our Com[missione]r – had been very much worried by the Lords of the Treasury.

April 4. At 7 p.m. Mr. A. Brassey addressed us at the Beaconsfield Hall, Shipton: as our future M.P. for North Oxon – the Honble. Bathurst also spoke well – one of the nicest political meetings I ever attended: stayed at the 'Rose & Crown' afterwards: till 12.30 with Messrs. Guy, Toy, Rose, Rowell, Breeder, Saunders, &c – from Chipping Norton.

April 17. Attended Chipg. Norton Board Guardians, & other meetings – Revd. T. Harris – resigned Chairmanship Board Guardians; Mr. Guy, & J.G. Abraham, elected – Mr. Neild & Addy, 17 to 9 Chairman & Vice-Chairman to District Council; Mr. Addy & Savidge, to Highway Board – and elected Mr. [left blank] from Norfolk, as District Surveyor of Highways, in place of Mr. Fredc. Forrester – we sat 4 hours, the largest attendance I ever saw!!!

April 18. By 12.15 train to Morton Steeple Chases, a great many attended from Birmingham, Dudley &c – and a very rough sworn lot of People they are!!!

I saw 5 out of 6 Events – and very close they all were, and appeared to *all* do their best to win.

April 20. Hounds met at Cornbury to finish the season – ran in Forest for an hour – then found in Singe Wood, & ran past Chasewoods House, & thro' Mr. Bolton's spinnies to Finstock top, dropt into Forest, ran the whole length of it to far corners of Ranger's Lawn, & down the valley between Mr. Sargent & Iorn's Farms, over the River & Railway & up the hill thro' the Jubilee Larches & finally to ground in 'Norell's'.

April 23. Planted Daffodils around the House &c – from Fairspear – they are fine *Old* Golden Flowers.

Planted French Beans. . . .

On Saturday Viscount Valentia C. polled 3754 votes for Oxford City – Dr J. Fletcher Little, Liberal 3143 – 602 Majority.

So in less than *3 weeks* Visct. Valentia, has *most unexpectedly*!!! become an M.P. thro' the *sudden* Death, of Sir Geo: Chesney.

April 27. Frequent Showers – mild. . . .

John Francis Fortescue Horner, appointed Comr. Woods &c: in the stead of Sir Nigel Kingscote – to our regret!!!

Municipal Buildings, Corn Exchange & Offices, on Gloucester Green, rapidly progressing, at Oxford.

May 15. Oxford County Agl. Show – Mr. Albert Brassey President, The Duke Marlborough, Earl Jersey, Visct. Valentia, Hermon Hodge M.P. – a large successful meeting: Robt. Willm. & Isabel attended.

May 16. At Witney sold Red Wheat 24s., Straw 28s. per ton.

Mr. F. Gask told me Mrs. Samuda, Bruern Abbey, had gone away with an Oxford Undergrad!! . . .

May 20. On Friday at Bristol, the famous Dr. W.G. Grace, made 288 v. Somerset – completing his 100 'Centuries', in first class Matches, during his 30 years *great* Career at Cricket – his *3rd greatest* score – in 1876 he made 318

not out – & 344 v. Kent – on 13th ult. he made 103 – as named, v. Sussex.
On 17th ult. – Notts. scored 726 viz., v. Sussex –
A.O. Jones 3 – Daft 46 – Gunn 219 – Flowers 64 – Dixon 35 – Attewell 4 – Howitt 119 – Bagguley 110 – Wilkinson 62 – Pike 23 – Handford 24 – Byes 12 – Wides 5 – Total 726!!!
This *great number*, is the *present record*, in an inter-County Match!!!

May 24. Her Majesty's 76th Birthday –
Met Mrs. Gask & Son – Mr. Fredc. Smith – at Fifield House – Shot 3 Dozen Rooks – 13 Dozens & 10 already shot.
The 'Duke', has Blue Eyes & light hair, got nicely over *Vaccination*: and is doing remarkably well.
Mr. John Jacobs, spent evening with Mr. Matthews, at the Manor House – slight rain at 7 a.m. – *but*, cleared off!!!

May 27. Alice 40 –
Farmed all round Icomb – ascended the Tower &c –
£19,000 lent, 4 per cent, on Mr. Sartoris's Estate.

May 31. Cold – with hot Sun – during the month, & scarcely any rain – checked Vegetation – very much.
The 25 acres recently planted plantation on High Lodge Farm – suffering very much – for want of rain.
Mr. Mrs. F. Matthews, safely reached Nuremburg.

June 6. The weather continues fearfully dry –
Drove Witney, Red Wheat 27s. Wool 20s. – Oats 16s.
Mr. Willm. Shillingford, & Thos. Taylor, Turk Dean, called and looked at the Wool.

June 7. Mr. Hayward Simpson, Icomb Place, 20 years old to-day.
Dined at George Inn, Leafield, at the Club Feast, Revd. Lee, & John Chas., there.

June 8. Leafield Feast passed off quietly this evening, so Whit *week*, has been extra Fine, for the multitudes of Holiday folks, all over the Country.

June 12. Saw Alderman Buckell, place the Gilded Ox, as weather Vane, on top of Oxford Municipal Buildings.

June 13. At Witney, Wheat 28s. – Oats 16s. – Wool 20s.6d. to 21s.
Dr. W.G. Grace, Clifton, nr. Bristol, is about to have a *large sum*, subscribed in 'shillings', to raise a national Memorial; he being by far the Best Cricketer, the World has ever known!!!

June 20. Her Majesty has reigned 58 years!!!

From Shipton to Reading 3s.9d. – hence to Ascot & back 3s.1d. & 3s.9d. Home.

Saw the Shahsada, 2nd Son of the Ameer of Afghanistan, his Uncle, 4 others, & his Page boy, in Royal Stand.

Called on J. Comins, the Royal Huntsman – who admitted me into Paddock, & Lord Ribblesdale's Stand –

Isinglass won the Gold Cup – he is the *largest* Race Horse, I ever saw – goes well in his Gallop – seems to canter – Loates pulled him 2nd two thirds of the 2½ miles, then took the lead, & kept it, 'Remainder' 2nd, 'Kilsallaghan' 3rd.

At Races, Ascot, saw Lord & Lady Yarborough, Chaplin, Heneage, & Dr. Fawsett, Louth!!

Saw 6 of the Rothschilds, 3 of them Ladies, fair & blue eyed, nothing of the 'Jewess' about them, in appearance!!!

Rose Brassey, & Capt. Scott – Stovin – Hadcock – Harvey, &c.

June 21. Finished carrying Hay & Clover at Langley: & Fairspear, to-morrow, all in excellent Condition & Quality.

Splendid weather, but sadly too Dry, for Farming.

Mrs. Matthews, her Duke, & Miss Rhys tea'd.

Isabel pulled the first 4 ripe Straw-berries, & Cherries.

June 24. Mr. McCalmont's 'Isinglass', ran his *last race*, when he won the Gold Cup on Thursday at Ascot – he has won *more money* in Stakes, than any other Horse on record!!!

The Emperor Willm. made a grand Display, in opening the Baltic Canal – on Thursday & Friday.[3]

Earl Rosebery, the Premier, resigned on Saturday evening!!![4] Marquis Salisbury, for the *3rd* time, summoned to Windsor, & asked by Her Majesty, to form a new Ministry!!! a very clever man.

On Saturday at Lords, Dr. W.G. Grace, made his *6th Century* of this Season – he & Sellers, batted 172 & not out, & won versus I Zingari[5] – 289 – 293 411 – 172.

Mr. John Addy, South Lawn, very ill from Sun stroke.

By rail from Shipton to Chipg. Norton, on returning joined Mr. Thos. Keck, & arranged for him to value John Chas. into Fairspear Farm on Saturday next: value, Furniture, Stock, Implements, & Acts of Husbandry, from 11th October 1894.[6]

An invite from Mr. Brassey, to attend Puppy Show, on 5th inst.

June 27. At Witney – Wheat 26s. – Oats 16s. Barley 18s. to 22s. – Beans 27s. – Wool 20s. to 22s.

Great complaints of the *continued* Drought – much Cropping quite half spoiled, rain, or no rain!!!

Consols 107¼!!![7]

July 5. By rail to Chipg. Norton – attended Mr. Brassey's Puppy Show, about 109 Lunched – Honble. Egerton Leigh, Col. Arnold, Capt. Becher, Revd. Lockwood, N. Long, Stephen Goodall &c – Mr. Cox from Baron Rothschild, Fredc. Travers from the Cotswold, & Edm. Bentley from the Whaddon Chase, were Judges – some very good Comic Songs were sung – I retd. with Bertee Davis, & R. Hambidge, to Ascott.

July 13. Posted my old friend Mr. Price 3 letters.

Mrs. Calvertt very unwell from a fall on Witney Flags, slipped up, awkwardly, on Orange peel.

July 14. . . . Yesterday, at 4.30 p.m. Mrs. John Charles Calvertt, Fairspear House, confined of a Daughter – my *first*, Grand-daughter!! Dorothy Esme.

July 20. . . . On Wednesday evening I settled all up with John Chas., he is *sole* Tenant for Fairspear Farm, from that date.

July 22. Our Polling day in North Oxon – Mr. Brassey 4057 Mr. Thornton 3074 – Majority 983.[8]

July 25. Drove to Witney – Wheat 24s. – Oats 15s. 16s. Barley 20s. 22s. Best tegs 7½d. lb. – Beef 6½d. lb.

Tea'd at Mr. Davenport's – Mr. Townsend & Shillingford called. Unprecedented!!! Lincolnshire Lustre Wool 32s. – Oxfordshire 22s.

Hermon Hodge, retd. for South Oxfordsh., 363 Majority.

July 28. Walked to Mr. Yapp's in 35 min. – accd. him to Leafield Church, on returning to Fairspear, called and had my *first* look, & kiss, of my Grandaughter the *First*; Dorothy Esme Calvertt.

Aug 1. Miss Edith & Bernard Abbott, came for the month.

Nice crops of Apples, Walnuts, Potatoes, Onions, Carrots &c, [there] are very fine prolific Samples at Langley – and in general, in Oxon.

Aug 4. Walked Fairspear, Leafield, not *5 acres* of Good White Corn, on Potter's Hill Farm.

Splendid growing Weather – for everything!!!

Aug 5. Bank Holiday.

My riding, driving, & hunting Horse 'Belfast' – broke his fore arm leg bone – while frisking about in the Paddock – in seeing 12 Beast & 6 Horses, driven into the yard – for separating: Steptoe, killed him – so we, and our Animals, are always in Danger!!!

The Revd. T. Harris, Swerford: 25 years Chairman to Chipg. Norton Board Guardians &c – died ae 84 – nice genial old Gentleman – on 2nd ult.

Aug 6. Mr. John Gregory, Enstone, died ae 69: thrown out of his Trap short time ago – internally injured: Insured against accidents for £1000 – Horse frightened thro' reins getting under his tail!!

Aug 10. Finished Wheat cutting at Langley.
Very delightful weather for *both* Harvest, & Vegetation!!!
By noon train to Oxford.
Called on Mr. Davis – who I found rather shaky.

Aug 13. Rained all the morning, will injure the early cut Corn, Potters Hill crops all cut, except Cone Wheat; and most of it, on High Lodge Farm – Meadow Hay around Oxford City – spoiling!!! Clover seed spoiling – the same will shortly apply to the Corn Crops in general.

Primrose League Fete in Shipton Court grounds next House – allowed by *Tenant* Mrs. Reid.

Continues too wet, for cutting Corn, or Horse-hoeing Turnips – very hindering – but exceptionally Growing. . . .

Aug 15. A *fine* Harvest morning, at last!!! *everybody*, cutting & tieing. Walked by 'Mayes Close', across Big Field, & Fairspear Lawn, by Chalk-pit spinney, & Fernhills valley to Forest Quarry in Witley Hill, on to High Lodge Walnut tree – and to the newly planted *dead* trees in Ascott Common, to Boynal Cover, & across Shakernook to Langley – never saw the crops so *light*, the past *19½ years* I have lived on the land.

Aug 19. My Grandaughter was christened at Icomb Church, on Friday last, her sponsors were, Misses Ellis & Clark, and Mr. Radford, & Hayward Simpson, her name Dorothy Esme.

Walked down & had an hour with Mr. Davis, Ascott Grange, admired his 6 years planted Victoria Plums & Tree – magnificent crops Apples in both his Orchards – Peaches – & Dahlias – very large, & fine.

Carried 16a. Barn Close Wheat, & 10 acres Shipton Piece Oats.

Aug 23. By 8.40 train to Cheltenham 10.10 – saw Sewell, & Townshend, play very good Cricket – rain 1.25 – left 3.20.
Cutting thin late Wheats on the Cotswold Hill – wretched late light crop.
Mr. Maddox's, Jeffery's & our Turnips, the best all the journey!!
Called & saw a horse at Mr. Hawker's, Junction, but did not buy him.
Rode on Matthews Coach to the Barrow – walked here in 30 min.

Aug 25. Walked and called on Mr. Yapp, on my way to Church – retd. by Fairspear, and called on Mrs. Dore, who I fear is on her death-bed!!!
John Chas., & Miss Ellis called in the evening.

Aug 30. 49 years to-day, since my Father died.
Shot 14 Woodpigeons, nr. the Hunting Gate, out of Langley, into South

Lawn Farm.

When does Truth, cease to be a Truth? When she lies at the bottom of a Well.

Why is a red-herring like an over-coat? Because it keeps you dry.

Why is a dog's tail like the heart of a Tree? Because it is farthest from the bark.

Sept 13. By rail to Chipg. Norton, & was driven to 'Cold Harbour' Stock Sale – Mr. Brassey has let the Farm 370a. – to a Mr. Saunders, 8s. per acre – Ewes averaged 48s. – Lambs 32s. – Beast cheap – bot. by Mr. Holtom, & Lusada.

Sept 14. Hot sunny days – with heavy Dew, at night – extra so.

78 Centuries, in first class Cricket, have been made this season, against 17 last season, & 43 in 1893.

Sept 18. Magnificent sunny day, with brisk South wind –

Finished carrying Clover for this year: cleared up everything, in Good! Condition!! Thrashing Vetches, Barley, Oats, & Wheat.

John Chas. & wife, gone to London – married a year to-morrow.

Pulling the *largest* Golden Pippins: Bon Creton pears ripe.

Sept 21. Yesterday finished thrashing for *this* year.

Drilling Wheat in Oatless – John Chas. *already* drilled Wheat in his 43 acres 'Farthings' – 12 acres Vetches – early, early, owing to the continued magnificent weather.

Sept 25. John Chas. drove Robt. Willm. & myself to late Mr. Brain's Stock & Implement Sale at Upper Slaughter: Mr. Taylor, Auctioneer: the Farm belongs the Brasseys, re-let at 8s. per acre to Mr. W.W. Hanks.

We tea'd Mrs. Matthews, Slaughter: reached Langley 7 p.m.

Sept 28. Sunny hot dry week – exceedingly so – so hot a Sept. was scarcely ever known.

Nov 4. The opening meet of the Heythrop Hounds – very much spoiled by Fog in the morning, & rain at noon, I lost nothing, by being absent.

Nov 5. Rainy dark night – Bonfires at a discount.

Nov 6. The Duke Marlborough married in New York, to Miss Consuelo Vanderbildt – a Gorgeous Wedding – hope it may be a success in every way, both for the Families, and Blenheim Estate!!![9]

Nov 13. My 66th Birthday.

Attended C. Norton Board Guardians – District Council – Rate 2s. in £ –

signed – Pay Parish 'calls' &c.

Hounds met Gawcombe – abundance Foxes, no scent.

Miss Reid, Shipton Court, broke her leg.

Nov 19. By 8.40 to Worcester, looked over Mr. Rd. Smith's Nursery – bot. 13 Roses, 4 Pyramid Plums, a green Bully tree[10] – a Quince – & a Mulberry for Langley.

Bot. Isabel a Worcester China, Breakfast Cup & Saucer!!

Nov 21. Isabel's 19th Birthday.

Planted a Mulberry, Quince, Green Bully, & 4 Pyramid Plums, & 13 Roses – all from Mr. R. Smith's, Worcester.

The 8 Roses around the Mulberry Tree, begin at *one*, opposite the Drawing Room, & walk to the left – names are Duke Connaught – Madme. Annie Wood – Mareschal Valliant – Marie Baunian – Duke Wellington – Abel Courier – Madme. Victor Verdier, & Alfred Colomb.

The 3 Roses opposite the Dining Room Window, begin at the one nearest South Lawn, names are Duke of Teck – Gloire Margottin – Dupuy Jamain –

The 2 Roses next centre Garden Door are – Duke Edinboro' nr. South Lawn, & General Jacqueninot –

4 Plums, 'Czar', *nearest* House, Old Orleans, Diamond & Victoria.

Dec 1. Splendidly fine calm morning.

Cut 3 Cauliflowers, grown in the rain & fog, the past week!!!

Walked around High Lodge Farm – the trees *all* planted a second time in Ascott Common.

Dec 7. Isabel, & Miss Hunt, walked to the Wilcote 'meet' and had Lunch with Mrs. Gask, at Bridewell.

Wm. Wells, Charlbury: gave us 50s. each for 30 tegs.

Mr. Edmonds pays 18s. per acre rent, for the Chilson Farm under Sq. Capel – it ruined Mr. Chas. Sargent – lost £8–10,000 there!!!

Cleaning the whole of Blenheim Lake, at a cost of £17,000!!!

Dec 14. Fog & Rain – wrote Mr. Price, Donington Road, Newbury.

Mr. Dark, Lew Hill, Witney – failed – debts £10,000 – been speculating on the Stock Exchange!!!

Dec 18. John Longshaw, Langley died ae 80, he worked in Wychwood Forest all his Life – he was a thorough 'Forester', after the deforestation in 1858 – he continued on Potters Hill Farm: the *changes* that have occurred around here, during his lifetime, are wonderful, to relate!!!

Dec 21. Had 3 hours at Oxford – splendid display of Meat, Poultry, Game, Fish, & even Flowers.

The Municipal Buildings *exterior*, finished: to view. The new Corn Exchange, & other buildings, also nearly finished. . . .

Dec 25. Mr. F.W.P. Matthews, Grace, & Son: dined.
Frost and an inch snow, last night, a cool white robe, for Xmas day.
On Monday & Tuesday, the 2nd Whip, 'Huntsman', and 2nd Horseman, 'Whips': Stovin ill – 'Will', broken collar bone, Mr. Brassey, lame eye: this *frost & snow*, ought to suit them *exactly*!!!

Dec 28. Foggy, thawy day – Frost, & inch snow, gone: mild.
Wrote Stovin & Dale!! both ill, of Influenza.

Dec 31. The past wretched, ruinous year, (for Owners, & Occupiers) ends. We are fighting the Ashantis, France, the Cubans – Italy, the Abyssinians; & Tom Fool, Brag, 'President Cleveland', will blow us all out of the Water – sometime during our Future eventful year 1896.[11]

1896

Jan 5. My Grandson, Marmaduke Matthews, a year old to-day. . . .

Jan 18. Hounds met Shipton Court – had a nice day.

Attended Oxford Corn Market in New Exchange, George St. – this was the 3rd time – '52 Stands', & 51 taken!!! it appears quite a success – a fine large, light room.

Jan 20. At 9 p.m. – ae 37 – Prince Henry of Battenburg, died on board the Cruiser 'Blonde', on the Coast of Sierra Leone, caught Fever on the way through Ashanti to Coomassie, & was returning to Madeira – *sudden* relapse – & died!!! Princess Beatrice – a Widow, 3 Sons & daughter.

Jan 31. Wrote Messrs. Clutton, & sent ¼ rent – only.

January, been exceedingly mild.

Feb 4. Attended Concert, Beaconsfield Hall – 3 Misses Reid, J. Reade, Schofield, Mrs. Grisewood, Coleman, Mrs. J. Chas. &c sang – all passed off most satisfactory, I retired to Crown till 11.35.

Wrote Mrs. A.E. Abbott, Nehim House, Bhusawal, Khandeish, Bombay Presidency, East India.

Feb 6. At Witney – brought away 4 shiny Crown pieces, for my Grandson Marmaduke Matthews – 'toys', he cannot wear out!!!

Wheat 28s. – Barley 18s. to 22s. – Oats 15s. to 16s.6d. Beans 27s. – Winter Vetches 40s. – Wheat Straw 35s. to 42s. – 9 st. tegs 8d. lb. – Beef 6d. lb.

Feb 11. Extraordinary fine mild weather, for everything. . . .

President Kruger, President of the Transvaal, has accepted our Foreign Secretary, Josh. Chamberlain's invitation, to visit London: & arrange South African affairs!!![1]

Thief Dark's, Stock, Crop, & Implement Sale, at Lew.

Parliament meets to-day: Marq. Salisbury, has upwards of 150 majority!!!

Feb 13. Breakfast at Fairspear – drove and called on Mrs. Adams, Hailey, on my way to Witney – calm trade – nothing doing.

. . . Grace Mrs. F.W.P. Matthews, confined of a Daughter, 11.30 p.m.

Feb 14. Hounds met Little Barrington – Mr. Matthews, horse came down, head over heels, in Bradwell Park – no bones broken – much bruised only!!!

Feb 15. L.M. Wynne's Bankruptcy –
Gross Debts £337,762 from 1886 to March 1895, £203,408 had been *withdrawn*, in *addition* to which £56,850 represented his losses on the Stock Exchange!!! . . .

Feb 17. Drove to the 'Bear Hotel', Woodstock: and walked over the Lake works – clearing out 40a. of the 70a. next Woodstock.
Had a look at Fair Rosamund's Well – and the stone enclosed Bath – the Water is extra cold, all the year round.
The Palace covers 3 acres of Ground, and much of the Stone was hauled from *Dog Kennels Quarry*, in Cornbury Forest.

Feb 18. Drove to Fifield – saw Grace & her 4½ days old Daughter, also Mr. Matthews, who is progressing favourably from his *fall* on Friday last.
Hounds met at 'Merry Mouth', but too foggy to hunt. . . .

Feb 28. Walked round High Lodge Farm.
Hounds ran a Fox from Dean Grove, & killed at Wootton.
February been a fine, mild, dry Month: Springs extra low – Thermometer up to 50!!!
Mr. Cockerel, called yesterday, from the Crown Office – said Mr. Clutton was in his 88th year, & ill of Bronchitis.

March 4. Attended School-meeting – to prepare for H.M. Inspector – & Messrs. Ferriman & Young met me, at Revd. T.W. Lee's.
Miss Matthews, came to visit, Isabel.
On Sunday Willm. Sampson, Langley, our Labourer for 20 years, walked to Charlbury for a bottle of Medicine for his wife, ate his Supper, & well apparently as usual – at 11.30 p.m., he gave a curious groan – & died ae 58 – exactly same as the late Duke Marlborough – failure of the Heart.

March 6. Mr. John Clutton, died of Bronchitis in London on Monday, in his 88th year – they have the [largest] Land &c Agent Firm, in England!!! . . .

March 8. Fine, mild, West-Windy day – good enough for April – Dorothy growing a fine sharp-looking, Grand-daughter!!

March 9. Shipton Auction, tea'd with Mr. Brookes, & attended Parish Meeting – a Poll demanded, 7 Councillors required, 14 *proposed*.

March 14. Settled up with the Crown – £50 allowed off the past two years rent on Langley.

Hounds met at Cornbury – I walked by 5 Ash bottom, and returned over the 'plain', up the valley to High Lodge.

Mr. Temple, showed me his Peach trees in full bloom under glass – the Garden looks superb – the rhododendrons on Lawn in front Cornbury House, are very full of bloom knobs – the place looks so neat & cosy – you regret it is not occupied.

Mr. Brassey told me his injured eye, was useless, for reading, writing, &c – Captn. Daly told me Ld. Willoughby de Broke, was riding extra bold & well: & having superb sport.

March 15. John Sargent died ae 39. Pneumonia.

March 16. Mrs. John Arthur Rigge, Northleigh, Henley on Thames, died from rupturing a blood vessel yesterday, ae 35 – very sad!!! She is only sister to Mr. Josh. Reade, Shipton.

March 30. Mr. Jefferys, J. Reade, J.F. Maddox, Hartley, Coombes, W. Cross, W. Longshaw, were returned as Councillors for Shipton – I voted for 6 of the above 7. . . .

March 31. The Duke Marlboro' & his Bride, came Home to Blenheim, met at Woodstock Stan. by Tenantry, Fire-brigade – Yeomanry, Town Council &c – Fireworks in Evening. Thousands welcomed them Home in beautiful weather.

April 1. Accompanied John Chas. to Board Guardians, C. Norton 33 attended – 17 voted for Compulsory vaccination, 11 against, & 5 Neutral!!! Mr. Mrs. Webb, the Master & Matron, gave notice to leave in a month, after 17 years services.[2]

April 8. Vaccination been neglected at Gloucester for some years, Small Pox very virulent – quite a Plague-smitten Town – shunned by all who can evade going to the place for any puropse – quite an object lesson for the Country.

April 10. Mr. John Joslin, Coombe; died on Wednesday from Heart Disease: Mr. J. migrated out of Devonshire, with several others on the Blenheim Estates, 10 years ago.

April 13. Robt. Wm. drove me to Icomb Place – climbed the high trees, & fastened pink flags for the 2 Point to Point Steeplechases, Gentleman's won by Capt. Wilde – Farmers by Mr. Hiorns.

Lunched in large Tent – Tea'd Mr. Simpson's – saw Dorothy there, for *first* time – did not know her *Mother.*

April 18. The Cuckoo has arrived here – had a magnificent week. *All* kinds of Grain, & Garden Produce, going on, most satisfactory.

Buried Mrs. Thos. Dores at Leafield, near Sampson, John Longshaw was buried at Shipton – making 3 deaths in this very small Parish in a few months!!!

April 26. At Leafield Church – called Fairspear.

Mr. Mrs. Abbott are spending 3 months at Bangalore, about 250 miles off Madras: 180,000 inhabitants.

May 14. Not less than a *Dozen* Nightingales, in Hensgrove Scrubbs – abundance Swallows & Turtle Doves – does this forbode an extra Dry, Sunny, Summer? it has that appearance at present!!!

May 18. A continuance of extraordinary Dry, Sunny, Weather – The Garden, & autumn planted trees, require frequent watering.

May 19. Cornbury Estate, advertised to be sold in July, in Oxford Journal.

May 22. John Clutton, Woodhatch, Reigate: died on March 1st ult – ae 86 – Personalty £148,283. The Crown Steward, and Receiver, for many years.

At Oxford, saw Somerset, badly beaten by Oxford, at Cricket – Witnessed the Performance of 'Trilby' –

Attended the Ch: Ch: Service, so did the Bishop of Reading.

May 26. Mrs. Hetty Ellis Alder, & Daughter, left Shipton, to join her Husband in America.

May 27. Drove to Breakfast at Fifield, & reached Icomb Place, 9.45, farmed round with Mr. S.M. Simpson, rain much required.

Bruern Farm, looks 'wretched': if I were Hodges, I should 'ticket', for Timbuctoo, & stay there!!!

May 29. Dined with the Leafield Club at the George Inn. Revd. T.W. Lee, in Chair – I opposite, carving a large piece of Beef for about 70 – Weather superb – for Holidaying!!

Smith from Scrivelsby, dined with us, & sang Red, White, & Blue.

An exceedingly Dry month – all over England.

June 7. Hayward Simpson, Icomb Place, 21 years old, inherits his fine Fortune, in 2 years time.

A really beautiful rain at last, *here*: much has fallen all over the Country the past week, especially in Linc. – after 6 months *extra* dry weather. . . .

June 16. By 8.10 reached Ascott 11 a.m. – called on Mr. J. Comins, at the Royal Kennels – he showed me his entry – 16½ couples – gave me tickets for Ld. Coventry's Stand, & Paddock – Shaddock & Quarrel, are two very good looking racers – each won.

Leonard Brassey's 'Pride', won the Vase.

Spent the night with Mr. Price, Donnington Road, Newbury.

June 17. Left Newbury 8.44 ran to Reading in 23 min. – arrived Ascot 11.15. Enjoyed the day's racing very much – Lunch with Mr. Mrs. Comins &c – in Ld. Coventry's Stand room – Champagne, Hock, Salmon, Cherry Pie, & Grapes: left 5.10 but did not reach Oxford till 7.10 left 8.38 Ascott 9.10 p.m.

June 18. Drove to Witney, news that Mr. Donald Currie's liner 'Drummond Castle', had struck a rock nr. Ushant, 264 drowned, 3 saved – sank in 3 min. – awful!!!

June 20. My Grand-daughter, Dorothy Calvertt, came to 4 p.m. Tea: cut her *first* rose off my newly planted 13 Standard roses: & cherries.

Queen Victoria completes 59 years reign to-day: and beats the record on 25th. Sept next!!!

June 24. Finished drilling Turnips at Langley. . . .

Mr. Bliss and his only Son Spencer drove here for Tea, Cherries, Strawberries, & Farming.

Robt. Willm. engaged to Miss North.

June 25. The first of 3 Test Matches, between England and Australia, at Lords –

Australia 53 & 347 – 400

England 292 & 111 – 403 & 6 wickets in hand!!!

For international Matches, decided in this Country, the record now stands at – England *eleven* wins, Australia *two*, & matches drawn *four*.

July 1. Cool, black-windy morning, threatening clouds, but the rain passes to the North.

Dr. W.G. Grace made 51 & 102, & carried his bat out – against Lancashire at Bristol, on Saturday.

July 8. Yesterday's Thunder & Lightning, smashed an Oak in Fairspear Lawn, cut half the Walnut tree down near High Lodge, & went down Mr. John G. Abraham's House chimney at Chadlington, & did much injury to interior of House.

July 9. At Witney – Tea'd Mr. Davenport's – drove round by Fairspear – settled with Mrs. Calvertt for Cornbury Strawberries: she was confined of a

Son, at 11.55 p.m. – so now she possesses a black daughter, & fair Son!!! in imitation of Grace, at Fifield!!!

July 15. 'St. Swithins', but no rain – dry as ever.

July 18. Harvest pretty general by Monday – all round; John Chas. reapering Wheat in Farthings.

Isabel at Northmoor.

Thousands acres Spring Corn in Oxon – will not be worth Seed & Labour.

July 21. Attended South Lawn Corn-Crops Auction, belonging Mr. John Addy, sold by Mr. Keck – 3 pieces Wheat only fetched £3 per acre and 1 piece Barley 57s.6d. – rest bot. in light & backward!!

July 23. New Wheat at Witney 21s. to 24s.6d. per qr.!!!

Began cutting Langley Farm 50a. Wheat on Monday, finished to-day.

July 31. Mr. Harvey du Cros, has bought the Cornbury Estate, of Lord Churchill – so we may expect great alterations: he is a Millionaire!!!

Aug 2. At Leafield Church – called on Mr. Yapp.

Walked down to Ascott, found Mr. Davis had gone to Brighton, and would hence, to Shepherd's Bush; and not return: I am sorry to lose so cheery a Neighbour, they were afraid of Paralysis, & Rheumatics, on the Evenlode.

Aug 6. At Witney, *Best* new Wheats 22s.!!!

Dr. W.G. Grace has performed wonderfully well once more, & beat all his scores since 1876 – at Bristol Cricket Ground, by making 301, the 9th out. . . .

Aug 7. Grace, & her fine boy, drove here, from Fifield.

On Wednesday the Duke & Duchess Marlborough gave a Garden Party at Blenheim Palace, over 700 of the County Gentry attended.

Aug 10. Mr. Harvey du Cros, will expend £20,000 over Cornbury House, in Decorations, Electric Light &c!!! exactly the Gentleman for the place.

Aug 12. Attended Oxford Ram Auction, bot. one of Mr. Gillett's, Cote: a very fine show of Sheep, & a good fair trade.

Lunch with Mr. Weaving – left 11.30 with Ram, to Shipton. Saw the 4 Brothers Stephens!!! one had been 20 years in New Zealand, & seven in Australia.

Carfax Church half cleared away, re-decorating the 'Clarendon', continually improving pretty Oxford – a beautiful city if you look over all its area.

Aug 21. To Cheltenham for Cricket – only saw Grace make two runs, & Rice, one: showers spoiled my day. . . .

On Wednesday last John Chas. Calvertt's, 1st son, christened, Maurice Calvertt, at Icomb Church: I have not yet seen him.

Aug 26. Drove to Bampton Fair – Slow Horse trade.

Mr. Harvey du Cros, *owner* of *Cornbury*, has a family of 6 Sons, the youngest not 21.

Aug 27. Drove to Witney – tea'd with Mr. Westell, born in 1815!!! Wheats 22s. to 24s.!! Barleys 18s. to 22s. – Oats 13s. to 15s.

Sept 1. Rain – altogether had the finest rains, since Saturday night, of the past *6 months.*

Sept 12. Met W.E. Morehouse, at C. Norton, with Langley Voters' list, only 10 – 2 Farmers, & 8 Agl. Labourers – viz – John S. Calvertt – Rt. W. Calvertt – John Hadland – Henry & C. Ferriman, Thos. Dores, Edm. Jacques – Stephen Moreton, Geo. Steed, & W. Steptoe.

Sept 13. Had frequent rains the past fortnight – the late Barleys all spoiled for Malting – suited everything else – very mild, & *growing*!!!

Sept 15. Mrs. Abbott came from Town after being away 4 years.

Sept 17. Sold 45 qr. Wheat to Mr. Holtom at Witney 25s. per qr. a ruinous price – extra good in quality.

Mr. Edmonds, Banbury, bid Mr. Elms £4500 for the 'Marlborough Hotel', clear of Furniture & Drinkables: Witney.

Sept 18. The Prince Wales is to spend a week at Blenheim in Novr. Shooting &c.

During the past 4 years, my daughter Mrs. Abbott has resided at Bhusawal 300 miles beyond Bombay, up country; she was 3 months at Bangalore, during the *Hot* Season – visited Bombay, & altogether, has seen very much of the World for years.

Sept 21. Drove to Chipg. Norton, & met Messrs. Guy, Neild, Abraham, & Surveyor Taxes, to re-assess, for the lately past Agl. Rating Act[3] – *very* rainy drive over the *Bleak* roads back, to Langley!!! many acres Barley & Oats, spoiling in the fields.

Sept 23. *This day*, Her Majesty Queen Victoria, beat the record, by having reigned longer than any, of her pre-decessors!!!

The Tsar, Tsarina, & their only child, the Grand-duchess Olga, are at Balmoral for 10 days, or a fortnight, on a visit to the Empresses Grandmother, our Beloved Queen!!!

Telegrams & Messages from all parts of the World, were forwarded to Balmoral to-day: expect Grand Rejoicings next June, when her Majesty, completes her *60* years reign!!! *All* our Royalties in England &c – are now with the Tsar, & Tsarina, the Premier, Marq. Salisbury is also there!!! . . .

Oct 6. Mr. Addy's Beast, Sheep, & Pigs; were sold at Shipton Auction yard by Mr. Bond & Keck: Beast, fair prices, tupped Ewes 37s. to 53s. – She Lambs 42s. – Ram Lambs 35s. to 64s. – He lambs 27s. to 37s. – one of the *best* Stock Sales, held this autumn.

Wind & rain, all the afternoon.

Oct 12. Attended Shipton Auction – sold Sow, & 12 pigs, 8 weeks old, for £4.17.0d.

Hired Stephen Moreton, Henry Ferriman, & Chas. Ferriman, as Carter, Shepherd, & General Servant.

Oct 20. Attended Mr. Abraham's funeral, at Leafield Baptist Chapel yard.[4]

Nov 2. Drove thro' C. Norton to Heythrop 'Opening Meet' – all the Farmers entertained at Luncheon. . . .

Nov 12. Wheat 35s. at Witney – Beautifully mild, still, clear Weather!! Cutting down Egg-Barn under-cover, & the large Ash, Elm & Oaks (in April) next – clear it out for years.

Capt. Evans, Kingham, died of Paralysis on Saturday, ae 72.

Nov 13. My 67th anniversary – hale and well – rather deaf!!

Nov 15. Drove to Dine with Mr. Eagle at Stanton Harcourt for the last time, he flits to Standlake in March – Mr. Bulford, Maule, & Hind, & Hatt, & Richd. joined us – Miss Walker – looked in, at Albert!!!

Nov 16. Albert Eagle goes to a Farm 407 acres, 140 arable only – at Somerton, 9 miles off Banbury, & 16 miles from Oxford, belongs Earl Jersey.

Nov 21. Three Misses F. Matthews drove here.

Grace, called with her Son, and Daughter; Donald Marmaduke walked up from Garden Door, to Front Door, for the *first* time, and did it in grand style!!!

Nov 26. Robt. Willm. drove me to the 'Bear', Woodstock: where I met Mr. G. Lamb, Fawler: we met the Duke Marlborough, Prince Wales, Lords Londonderry, Gosford, Chesterfield, Mr. Chaplin, Grenfell, &c – at the Stone Bridge opposite the Palace, and commenced rabbit shooting by

Rosamund's Well, they shot 2075 Hares & Rabbits before Lunch, and 1254 after!!!

Tea'd at the 'Bear', and walked out at 8 p.m. to see illuminated Cycles – the Column was illuminated – Bonfires – a *grand* display of Fireworks by Brock, after which 1000 Torch bearers walked from Column to North Front of Palace, as a *Finish* at 11.30 – reached *here* at 2.5 a.m.

Brought away a pocketful of the Prince's brass Cartridges as Souvenirs of the greatest days Shooting I ever saw: they used 3 Guns each, & kept them frequently all at work!!! The Princess Wales, Princess Victoria, Prince & Princess Charles of Denmark, & the Duchess, spent the day at Oxford – Countess Curzon, Londonderry, & 5 other Ladies including Lady Randolph Churchill, Lunched, & spent the afternoon with us Shooting – reached Langley at 2.5 a.m. The Royal Party arrived at Blenheim on Monday, & left on Saturday, had fine, dry Weather, all the week, shot 4 days. . . .

Dec 5. By 8.15 train to Oxford, left 11.12, reached Chippenham at 12.50, met, and driven to Badminton 10½ miles.

Mr. Wemyss had hunted, & killed lease Foxes, a lease were killed by Marq. Worcester, yesterday: his Lordship hunts the Dog pack & Mr. Wemyss the Bitches & mixed packs – and Willm. Dale, assists both – very much!!! he will fill Mr. Wemyss' place next Season: & no doubt eventually hunt the whole 5 days per week.

Dec 6. Attended Badminton Church – Beautiful monuments to the Somerset Family – the 2nd Duke died ae 30 only, the 3rd Duke ae 37 – succeeded by his Brother!!!

Lord & Lady Worcester – joined us in the Park, exercising Hounds at 4 p.m. Mr. Wemyss, Princess, a Count, & 3 other ladies, joined us in the Kennels.

Dec 7. Was driven thro' Tetbury to 'Trouble House' 10 miles – Mr. Hamblin joined us at Tetbury, situated on very hilly ground, with a Church with *large* windows *all* round.

Mr. Wemyss hunted the Hounds – bad scent – for a wonder – but storms coming.

Walked into Mr. Wemyss House in the Evening: Whiskey.

Dec 8. Left at 9.45 for Chippenham, admired the Ancient stone mile stones, on the Bath Road; one was 99 miles from Hyde Park Corner!!! Oxford reached at 2 p.m. – where the Builders at Carfax Corner, & Castle Corner, were all at work – as if in June – left 3.40 for Shipton: 4.23.

Dec 14. Building all up next Carfax Tower, all hands on, like June: also the block of new buildings, adjoining the *Castle* Hotel.

Dec 22. Walked from 5 Ash valley, thro' Cornbury Forest, to the Ponds; the old 'Newell' Pond, has been enlarged, & 2 new ones added, the stream cleared out down to the gravel; from these 3 Ponds, to the large Pond, by the Dog-Kennels – when completed it will be a very pretty Pond-valley, *all* along to the late Keeper *F. Shayler's House.*

Mr. Du Cros, is doing great improvements to the House and Estate generally it – will become a *Gem* of a Place.

Dec 31. This has been the *3rd* Dry Season, of the past 4 years!!

Land owners, & Occupiers, have been the most suffering class in England, the past 20 years!!! many ruined, & *all* lost Capital – a gleam of hope is now setting in – School Grants – Light Railways – half Rates allowed on Agl. land – & a rise in Corn – Wheat up to 31s. – & a reduction by the Railways, on Agl. Produce!!![5]

1897

Jan 1. A calm, bright, cheerful morning – in which to begin this *extra*-eventful year – the Queen has been Empress of India 20 years, & on 20th next June, will have been Queen 60 years!!! – of England & its *immence* Foreign Possessions!!!

Jan 7. Rode with John Chas. to Witney new year's market – *Fat* tegs 37s.6d. to 44s. *only* – Red Wheat 32s. – *very* little Corn offering.

Sent my Grandson, Bernard Abbott, a Hamper, Papers, Cards, Apples, Oranges, Lemons, Figs, Dates, &c.

Hawkins told me this will be their Record Game shooting season, over their Eynsham preserves.

Jas. Fras. Mason, fired off 991 Cartridges, *one* day, at Blenheim!!!! after the Prince of Wales's visit!!!

Jan 10. Walked down to Mr. Jeffrey's, Shipton in the afternoon, he told me, he should quit his Farm, under Dr. Jay, & Akerman, 29th Sept. next.

Jan 11. Attended Shipton Auction, average tegs 38s. to 44s. – a many tegs only made 27s. to 35s.!!! explained by the great scarcity, of Turnips.

Jan 13. Already we hear of great preparations being made for this extraordinary Jubilee Summer – *all* are hoping, the Queen will enjoy the best of Health.

Jan 14. The Crown allow me 30 per cent return, upon the Langley Farm to Michs. 1896 – £275 – £90 off – *32½ per cent*!!

Walked over to Fairspear, found John Chas. allowed £150 off, £450.

Mr. Mrs. Simpson, Mrs. F. Matthews & 3 daurs., Mrs. Matthews & Maria: Wm. Matthews, Mr. Mrs. & Miss Gask – Mr. Akers & Sister – Miss North[1] – Messrs. Cheatles, F. Smith &c, spent the evening at Fairspear.

Jan 16. Walked to the new Ponds, from Langley, down 5 Ash Valley in an hour; met with Grant the Fish-keeper, and walked all down the valley to Mr. Proctor's House, saw the Trout eggs in hatching House, Waterworks, & 55 men cleaning out the old Duck Ponds – walked up to Cornbury House, found 4 sets of Contractors at work – Building, Decorating, Drainage, Painting, Electro lighting, &c: on to the Gardens, where they are placing down 4 Glass Houses of Teak, iron, & glass – for Vines & Plants – &c; this Contract is £3000 – Cornbury will be a very complete Country House.

Jan 17. Mr. Cobb & Miss North, dined: after Church!!!
Half an inch snow, with Frost – still – very seasonable.
Robt. Willm. drove Miss North to Fifield.
The least Game on the Langley Farm, during the 4 past years.
Dr. Piggott late of Burford Grammar School, died of Heart Disease. . . .

Jan 22. Hawkins sent me 2 br. Pheasants. Thrashing Wheat, Oats, & Barley. North Wind, 1 to 2 inches snow, with Frost.

Mr. Joseph Ray Ensor Davis, late of Ascott Grange, & lately of Bucknall House, Shepherd's Bush Green – died last Monday ae 63 – so did Mr. C.D. Beauchamp. . . .[2]

Jan 31. . . . Two inches snow on the level – no wind – not very severe frost – in fact a very nice fortnight's storm already.

Mr. Freeman-Mitford, Batsford: very much reduced in 'Personalty', fooled it away, 'magnificently'!!

£200,000 already subscribed at Mansion House, to the Indian Famine Fund!!!

Feb 5. Had a few wretched days bad weather; fog, frost, snow, & the *greatest* rain the past night, & to-day – for the last 2 years!!!

The Plague, *awfully alarming*, along with the *Famine*, in India – especially in Bombay, & Karachi, – the Plague!!

Feb 9. The Indian Famine Fund now amounts to £238,000!!!
Consols 112½.

Mr. Woolliams, Ascott; called with a Petition, to G.W.R. – to allow 2 more of their trains to daily stop at Ascott Stan.

Feb 12. Mr. F.W.P. Matthews drove up for an hour –

Mr. Brassey fell on Monday, & slipped out, his injured knee, I fear it will soon be all over with his Hunting.

Feb 16. Walked to Cornbury Ponds, two Undergrads, had caught 7 Pike & a fine Perch 1½ lb. – the Pike 12 lbs., to 5 lbs.

Looked over the Works going on in the House & Gardens: and Ponds; 200 to 300 men at various work.

Examined the ice-house, in the Park.

Feb 24. Met Hounds at Bradwell – found in Sturt, ran to Grove, out past Bradwell House to Covers beyond, back to the Grove, and away to Kilkenny, turned & ran down & crossed Windrush at Widford, thro' Beech Grove to Widley, out to plank Quarries & lost entering Waterloo – found 2nd Fox on Rt. Jacobs's farm lost in 15 m. – tried Jolly's blank, but found in Westwell Copse lost in a cold bitter wind in 20 min – large field.

Feb 28. Robt. Willm. drove me to Fifield to dine with Mr. Fredc. & Grace.

February has been a mild month from the 3rd ult, not even had a rime frost – a *great* rain on 5th. – every appearance of a fairly forward Spring.

March 2. Met Hounds 'Merry Mouth', fair sport over Tangley.

Grace confined of her 2nd Son.

March 8. Shipton Auction – called Mr. Brookes – & Jeffreys –

Voly. School Meeting, at Revd. W.T. Lee's, Leafield Vicarage: rate 2½d. in £.

Weather continues nicely mild – with every prospect of a fairly forward Spring.

March 10. I see in the 'Field' –

There is only one Game at Cards!! thats Whist –
There is only one Wine!! thats Port –
There is only one Life? thats a Farmer's –
There is only one Sport!! thats Fox-hunting.

March 19. Mrs. Matthews, Fifield Manor House, seriously ill, from Measles.

Fine wild-windy, drying! day.

Walked down to Ascott, & bot. load Poles of Hanks 14s.

March 22. It is astonishing the money that is being given away for *Charitable* Purposes, this *Diamond Jubilee Year*: the Mansion House has already sent £443,000, Lancashire £100,000 and Glasgow £100,000 to relieve the *Indian Famine* – a splendid sum is being collected for the London Hospitals – and numerous other Charities.

Her Majesty is now at Cimiez, in the Best of Health – Sunday the 20th next June will be the *Great* Thanksgiving day, *all* over her Majesty's *Dominions*: the *Pageant* to St. Paul's, will take place on Tuesday the 22nd. June!!! – it will *surpass* anything, *ever* seen in London!!!

March 25. Mr. Jas. Shillingford told me he had been to London, to consult a Cancer Expert – who told him, he dare not operate – as his 'Fits', of Asthma Coughing, would cause death by Hæmmorhage, after the Operation: so poor man he has to linger on and die!!!

March 27. Mr. Simpson showed me a *large* stone, the size of an *average* sized hen's egg, taken out of Mr. Lynes' bladder – and a Watch, that was shot into a man's groin, thro' foolishly pulling his Gun by the muzzle thro' a hedge: also a small piece of Napoleon's Coffin!!!

Returned from Icomb – called, and saw Mrs. Grace Matthews, & her 2nd Son, (for the first time) the boy will be very much like his Father. . . .

Hunt meet, Market Square, Bampton *c.* 1900. *(Oxfordshire County Libraries)*

March 30. Isler & Co., have bored down for Water, in Mr. Yapp's kitchen, at Lowboro', 130 feet, and now expect abundance of Water at 138 feet. . . .

Immence preparations, are being made for the Festivities, &c, in June!!! – Many thousands of Pounds will be given for Windows & erections, along the *six miles* of Jubilee route, to St. Paul's, & back.

'Manifesto', won the Grand National: & 'Winkfield', the Linc. Handicap.

April 5. Voted for 7 Rural District Councillors at Shipton School House, viz Joseph Reade, Fredc. Jeffreys, Rd. Hartley, Wm. Cross, Saml. Coombes, Walter Longshaw, & Alfred Miles: the following are the Radicals, viz J.F. Maddox, A.F. Willis, Jas. Wright, J. Arundell, H.J. Coombes, Wm. Durham, G. Turner – *all* the Radicals returned!!! . . .

April 6. Seats fetching *fabulous* prices, along the 6 miles route, of the Royal Thanksgiving Procession, on 22nd June, from Buckingham Palace, to St. Paul's, & back –

Mr. Frank Matthews, bid £1000 for his site, in Piccadilly.

April 15. At Witney, saw Jas. Shillingford, who is suffering from a Cancer on the Tongue, very sad, painful case, as after much suffering & no hope, you die exhausted!!

Robt. Willm. gone into North Wales.

April 19. Robt. Willm. is now at Elan's Castle Hotel, Dolwyddelan, N. Wales: about 5 miles from Snowdon.

Walked to top of Moel Siabod, the 4th highest mountain in Wales, about 2600 feet high – snow on its top – scenery magnificent.

April 22. Mr. Freer, Kitebrook House, nr. Moreton: took his delicate wife to Jamaica, to benefit her health, he caught Fever, and died; on her returning and in sight of Southampton last Thursday, she fell over-board off the 'Atrato', & was drowned, leaving 3 orphan children – she was Sister of Mr. Geo. Dawkins, Wilcote, I knew them both – a sad, pitiable case.

May 1. Up & walked to Magdalen Tower – as soon as the Tower Clock struck 5 – the surpliced Choir, sang a Latin hymn which lasted 8 minutes, then 5 minutes on the Bells – a two minute interval – then they rang a general peal 15 minutes and all the *annual* ceremony was over.

By 9.18 train for Newbury at 11.19 – Dined at Mr. Price's, found him in better health than I expected – left at 3.53 reached Oxford 5.33. . . .

May 13. Walked down the Cornbury Fish Ponds, overlooked the works & men at House, building a central Tower, which will give the House a very conspicuous aspect, especially with an electric Ball-light at the top of it.

Had an hour with Mr. Lewis at Rangers Lodge – saw 180 hen Pheasants up for laying &c.

May 17. Jubilee doings all the rage, the daily Paper will be very interesting, the next 2 months: the Queen is in excellent Health, with every appearance, of keeping all her various Duties, & appointments. . . .

May 18. Had an Hour with Jas. Shillingford, yesterday morning, suffering from Cancer, under the Tongue, can scarcely move his Tongue, or talk: he wrote on paper: and I answered his Questions, & did all the talking.

May 19. The Princess of Wales Fund, to feed the Destitute Poor, on Jubilee day, has been presented with a £25,000 cheque, from an unknown Donor!!! it is *marvellous* the amount of money raised in a great variety of ways, thro' this Wonderful Diamond Jubilee, or 60 years reign!!!

May 21. Mr. Chandy [or] Chaundy of Ascott, found dead, in the Evenlode. . . .[3]

The Queen opens Sheffield Town Hall!!! on her way to Balmoral!!!

May 22. Fine sunny weather with N.E. & N.W. Winds, spoiling the Hay & Clover Crops.

Isabel, at Henley on Thames: with her Sister Mrs. Abbott, at 6, Norman Avenue.

May 24. Our magnificent Queen Victoria, 78 years old this 24th May 1897 – she is at Balmoral!!

May 27. At Witney – Wheat 27s. Wool 20s.!!! and *slow* demand.

Mr. Jas. Shillingford, gradually worse – cannot talk – or even be driven out!! Chair!! Tongue stationary: next to useless!!!

June 1. Isabel returned from visiting her sister at Henley on Thames.

Mr. Clappen, Draper, Witney: took 'vermin killer', on Saturday – and died ae 52 – a very nice, Business-man: depressed in spirits: similar to poor Mr. Chaundy of Ascott: very sad for relatives!!! & all concerned.

June 9. By 4.23 from Ascott, to Honeybourne, Halton, Warwick to Leamington 3s.9d. – to N. Western 9½d. to Coventry – 8 p.m. Stayed night at Mr. A. Baker's Temperance Hotel, Fleet St., Coventry.

June 10. At 9.30 met Mr. Radford, at Machinist Works & arranged about Lodgings for Fredc. Geo. – &c.

Coventry contains 86,000 inhabitants and is *now* doing a record Business, in the Manufacture of Cycles, Watches, and Ribbons!! *all* of which, are supposed to be the *Best*, in the World

Left at 12.35 reached Chipping Norton Junction 6.30 rode to Fifield & tea'd with Mr. F.W.P. Matthews – Miss Bull, there.

Witney Mkt., Wheat 27s. Wool 20s.

June 16. At 11 a.m. Robert Price, late of Shipton, now of The Firs, Berries Bank, Greenham, Newbury, Berks – died ae 74. My very good neighbour for many years, he excelled at Farming, Shooting, & Hunting: all his life – born a Welshman – his Family estates joined Lord Mostyn: Hope Hall, Flintshire.

June 19. Mr. Price buried at Greenham, Newbury: this afternoon.
Sixteen carriages will precede the Queen in the Royal Procession to St. Pauls – The Princess Wales, & Christian, will be in Her Majesty's Carriage – The Soldiers of *all* ranks, will number 44,303, riding 6333 Horses.

June 20. This morning, the Queen completed 60 years reign, extra Jubilee Prayers in all the Churches in England – the Queen attends St. George's Chapel, Windsor.

June 21. By 8.18 train from Ascott, reached Paddington 10.25 and Mrs. Abbott's, née Miss Pulford, 53 Guildford St. Russell Square at 11.
At 12.30 saw the Queen, driven thro' the Marble Arch for Buckingham Palace – and remarkably well she looked.
At the Alhambra – not so entertaining as 10 years ago!!!

June 22. Reached the 'Mocha' 336 Strand, at 8 a.m. – to witness the Queen's *wonderful* Diamond Jubilee Procession to St. Paul's – the grandest sight of my Life!!!
Mr. Abbott accompanied me, & retd. to Henley in p.m. Witnessed the *clever* Burgess Minstrels for an hour – Admired the *splendid* Illuminations from Trafalgar Sq: St. James' St., Piccadilly, walked from Hyde Park Corner to Marble Arch, & Oxford St., High Holborn, Southampton Row, to 53 Guildford St. 11.15: well satisfied with my wonderful day's Historical Performances.
The immence crowds behaved admirably – not heard of a Death, or even broken Bone!!!

June 23. Sent my Friends in Lincs. &c – about a score of Papers.
Walked to underground Railway at King's Cross in 10 min – rode to Baker St. Stan. & walked into Madme. Tussaud's – very pleased with all I saw. 'Bus' to Charing Cross – 'Bus' to Temple Bar Griffin – up Fetter Lane to Holborn & Guildford St. at 1 p.m. – left at 2 – walked to King's X – 'Underground' to Praed St. – left Paddn. 4.45 reached Ascott 7.20 after a *delightful* visit, in London.

June 27. Walked to Mr. P.J. Yapp's & accd. him, & Miss Metcalf – to Leafield Church.
Jas. Shillingford died ae 58.

June 29. On Saturday, the *greatest* grandest Naval Review at Spithead the World ever saw – 165 war ships – about half our Navy took part in it – 22 Foreign war ships were present – the Russian one, the most powerful.
Isabel was there.

June 30. June has been a very genial Productive month, and most suitable in Weather, for *all* Kinds of Vegetation!!! & Jubilee!!!

July 2. The Grass Mower cut the two Meadow fields, yesterday & to-day – Dorothy, & Maurice Calvertt, came to tea, Cherries & Strawberries: & exceedingly well they looked.
The Queen's Jubilee visitors are daily returning to their various countries, after the most extraordinary visit of their lives!!!!

July 12. At Shipton Auction – almost *too splendid* Weather!!!
Mr. Bliss, Chadlington, drove his 7 years old son . . . over here, & farmed round – Master Spencer, admired the Wild Ducks, & tasted the Currants, and Raspberries.
Dorothy Calvertt, Fairspear, is 2 years old *to-morrow*!!

July 27. Messrs. Hall & Co., Oxford; bot. Marlborough Hotel, Witney, of Mr. Geo. Elms, for £4500.
Spring Hall, Shipton: advertised for Sale 257a. 1r. 10p. – rent £294.4.8d. Rectorial Tithe £45.7.2d. present value £31.14.0d., Vicarial Tithe £13.13.1d., now £9.10.0d. Mr Jeffreys: tenant.[4]

July 31. Rent day – Mr. Nelson left.
Fredc. George, came over from Coventry, for a week's holiday.
The best Harvest, to cut: of the five, I have grown at Langley. John Chas. began Harvest at Fairspear. . . .
The Church of Wath on Dearne, Rotherham, Yorks. – The Revd. Herbert Ward is vicar – his wife my eldest Daughter. Adwick, is also attached to this Living.

Aug 9. Began cutting Wheat at 9 a.m. – & Tieing at noon.
Shipton Auction – Mr. Jeffreys taken the 'Meade Farm' Chipping Norton – owned by Mrs. Prestige.
Mr. Peter Yapp bids £5000 for Jeffreys Farm, with a £59 Tithe.

Aug 11. Attended Oxford Great Ram Fair, Auctions –
Mr. Brassey's Shearling Rams averaged £14.16.0d & his Lambs £6.7.0d. Mr. John Worley's £9.8.6d. Mr. G. Harris' Ram Lambs £7.6.0d.
At Winchenden, Mr. John Treadwell's 62 Shearling Rams, averaged £17.19.0d. – 18 to go to Germany, & 1 to Ireland.

Aug 13. At Cheltenham W.G. Grace the renowned Cricketer made 131 without giving a chance!!! the 115th time he has completed over 100 at 1st class cricket – he is in his 50th year. . . .

Aug 16. High Lodge Farm, advertised for re-letting, after being 'in-hand' for 5 years.

Aug 25th. Finished cutting my 45th Harvest!!!

On 1st of this month sound red old wheats worth 27s. – now worth 37s.!!! caused by bad weather, & bad reports from various Countries!!!

Aug 28. Weather showery, and most unsettled. . . .

Mr. Davenport cycled over here from Witney.

Messrs. Hall & Co. bot. the Marlborough Hotel of Mr. Geo. Elms, and have let it, to a Mr. Falkner.

Fredc. Geo: Cycled here from Coventry via Banbury, about 50 miles.

Aug 29. Robt. Willm. drove me to Fifield – Dined with Mr. F.P. Matthews. Showers – Corn cut, & uncut – is in a very critical state.

Aug 30. Mr. Jeffrey's Shipton Farm offered by auction on Tuesday, no bidders!! . . .

Sept 1. At Chipping Norton Board Guardians, & District Council – Much Corn to cut between here, and Banbury: spoiling in the fields thro' continued frequent showers.

The Hops, & hop-*Pickers*, in a wretched condition in Kent.

Sept 11. Finished carrying Harvest, except 6 loads Oats – only been able to carry 3½ days the past fortnight – thro' frequent rains, & Thunder-showers.

Sept 13. Very glad to finish carrying Harvest:

Witney Carnival, or Feast: no Business.

At Shipton Auction, 412 Sheep, 31 Beast – rough lot.

Sept 23. Drove to Witney – Mkt. very small, & irregular –

Mr. Richd. Eagle much crushed behind a Gate by a Beast rushing at the Gate.

On Saturday 18th. ult – the Duchess Marlborough confined of a Son & Heir, to the Blenheim Estates, & *much wealth*, from the American Vanderbildts!!!

Sept 25. Met Mr. Banister, Mirehouse: at Chipg. Norton Town Hall, with Langley Parish Voters' list –

Burford Fair –

Mr. John Jacobs, Corn Mill, burnt down.

Fredc. George, cycled home from Coventry, thro' Warwick, Brailes, Golden Cross, Moreton in Marsh, Evenlode, Adlestrop, Kingham, &c.

Sept 30. Josh. Reade bot. Mr. Jeffrey's Farm, which he is quitting at Shipton –

Mr. John Fowler Maddox, quits Shipton, Michs. 1898.

Nov 24. Reached Home after a pleasant visit to Stamford, Retford, Lincoln, & Broughton nr. Brigg.

Nov 28. At 8 a.m. taken [ill with] violent pain nr. Heart & shortness breath, bad enough to die for ½ an hour. . . .

Dec 1. Four days & three nights, could neither eat, drink, or sleep – Dr. Cheatle came every day. . . .

Dec 8. The Honble. Mr. Mrs. Hardinge & child, came into residence at South Lawn.

Mr. du Cros, gave £115,000 for Cornbury, *entire* and £35,000 *more*, was paid by *another gentleman*, the real price being £150,000. Oh! dear! Oh! dear!! £100,000 would have been quite a *Good Price* for it – so lucky Lord Churchill!! the *sale*, is much to his credit, all thro' the affair!!![5]

Dec 31. Gradually improved since Xmas Day.

Exceedingly fine, mild, open weather!!!

This past year has been a *great improvement* upon the *Previous* 7 years, in Farming: will always be memorable in History, as Victoria's Diamond Jubilee Year!!! – the grandest sight ever beheld in London. . . .

1898

Jan 8. Exceedingly fine mild weather since Xmas Day!!! Sufficiently recovered from Bronchitis & Pleurisy, so was driven and called upon Mrs. Adams, Hailey Manor, then forward & met Hounds at Eynsham Hall, shook hands with Mr. Mason, & Mr. Brassey and welcomed Mr. B. on his safe return from Melbourne, had Champagne &c – Fras. Hawkins gave me a brace Ck. Pheasants – returned by Witney, & called upon Mrs. Abraham, late of Minster Lovell, &c.

Jan 15. John Baughan, Shipton, a bankrupt.

Jan 24. Thos. Garne, appraiser, Burford, died of apoplexy, ae 45.

Jan 29. Met Hounds, Cornbury Park; inspected the new entrance on Park side of Cornbury House, outbuildings, Ranger's Lodge, Shayler's House, Gardens, Ponds, newly planted trees, & many other *great* improvements, all *about*, and *around* the Estate; will be a Gem of a Place, when *completed*!!!

Jan 31. Upwards of 100,000 Engineers & their Labourers commence work to-day after a ruinous 'Strike', of 30 weeks – expending their £300,000 Funds & experiencing a Labour Wage Loss of £2,000,000!!![1]

January, has been a fine, *mild*, dry month!

Feb 6. Walked down to Shipton and dined with my Daughter, Mrs. F.W.P. Matthews, in Tothill Villa, their new House, built last summer by Mr. Peter Yapp, Sloane St., London: outlay £1500.

Feb 28. February, has been a very mild, dry, month –

Red Wheats 36s. to 37s.6d. – *Ordinary* Barleys up to 25s. to 27s. – Oats 16s. to 18s.

March 1. Mr. Thos. Keck, Bourton on the Water, died ae 50 – he succeeded his Father, only 3 years ago, as Auctioneer, held auction Stock Sales, monthly, at Chipping Norton, Shipton, Charlbury, C. Norton Junction, & Hook Norton.

April 8. . . . Sirdar Kitchener, Commander in the Soudan: fought the Dervishes at Atbara, 3000 slain: Emir Mahmoud taken prisoner: at the next rise of the Nile, they intend to fight thro' Omdurman, to Khartoum, and revenge the death of Gordon.[2]

April 21. . . . Mr. Sutton, Wilcote: sold red wheat at Witney, 43s. per qr.
Mr. Wm. Dale writes me from Badminton, to say, their Hounds have been out 149 times, killed 160 Foxes, & ran 46 to ground.

April 28. At Manchester during the past week, wheats have risen 6s. per qr. – the 4 lb. loaf to 7d. – the *general* prices *just now*, are red wheat 46s. to 48s. – Barley *grinding* 28s. – Oats 18s. to 22s.
The Country is looking most productive, & every appearance of an early Harvest – Pastures, in full stock – Fruit trees, very full of Bloom – Farming has not looked so rosy, the past 20 years!!! such a 20 years depression, thro' *low* prices, & *wretched* seasons, have not been experienced in England since the Civil War of Chas. 1st. &c.

May 9. Mr. Davenport leaves Witney Metn. Bank, for Bromsgrove.
The 190 hen Pheasants at Rangers Lodge, average laying 120 to 130 eggs, daily.
Cherry trees in Bloom.

May 19. Mr. Wm. Ewart Gladstone died – born Decr. 29th 1809!!! the most clever, and hard-working Politician of the Victorian Era – was buried in Westminster Abbey.

May 30. Whitmonday Sports in Cornbury Park – great attendance.
Jumping & Driving Competitions, Athletic Sports, Band Music &c. Looked round the Gardens & much admired the greatly improved ancient Country House.

May 31. Had a rainy, showery, very fine Month, for all kinds of Vegetation.

June 7. Mr. Hayward Simpson, Icomb: 23 – inherits the Income of his late Uncle Dr. Hayward's Property, who resided, & practised a many years at Stow on the Wold – about £4000 per year.

June 11. The *Giant* financier, Mr. Terah Hooley, made a Bankrupt – will much affect Cornbury, & Messrs. *Du Cros*!!![3]. . . .

June 15. To Ascot Races, Jacquemart, won the Royal Hunt Cup, and Elfa II, won the Gold Cup.
Saw Prince Wales, Duke Cambridge, Lords Heneage, & Coventry, Sir Nigel Kingscote, A. Brassey, &c.

June 29. Mr. Commissioner Horner, Clutton, & Hellard: called here, while looking over the Crown Estate.
Mr. Richard Stovin, late Heythrop Huntsman for 8 years only, has received a parting souvenir of £716.9.6d. – they gave him over £800 when

he left the Bicester 8 years ago.

The Queen is in first rate health at Windsor, she has become the greatest monarch-wonder of the Age!!!

July 4. Maurice & Dorothy Calvertt came to Tea, and met their Cousins, Bernard & Beatrice Abbott who is a year old to-day!! . . .

July 27. Mr. Simpson, Sutton: nr. Epsom: farmed me round.

Wheat 36s. – Wool 18s. only!!! . . .

Been a very *Dry*, but not very sun-shiney month!!!

Aug 10. Oxford Ram Fair – fair trade – Mr. Gillett's averaged £10.9.8d. Mr. John Worley's averaged £10.10.0d. – Mr. Geo. Adams 70 lambs averaged £9.4.8d. Mr. Bryan's £8.8.0d., Mr. Treadwell, Winchendon, 62 averaged £17.14.0d. Mr. Brassey's Shearling Rams averaged £9.13.0d. Stilgoe £10.4.9d. Jeffrey's lambs £7.12.3d. – not much *Foreign* Competition: so were sold at *reasonable Breeders'* prices.

Aug 29. Yorkshire, the Champion Cricket County, played 26 Matches, won 16, lost 3 – drawn 7.

Gloucestershire, played 20 Matches, won 9, lost 3, drawn 8 – W.G. Grace 1141 runs – average batting 47.54!!!

Aug 31. Been a very fine month for Farming, the best Harvest of the past 22 years in Oxfordshire – is being gathered in excellent *Condition*.

Sept 2. The Battle of Omdurman, under Sir Herbert Kitchener, the Sirdar, was fought this day near Khartoum 11,200 Dervishes were counted dead on the Field, & 15,000 are said to have been wounded, the 21st Lancers distinguished themselves by cutting a road through 3000 infantry – made brave noble incidents during the fighting – cleared the East Soudan, for the future great benefit, of Egypt.[4]

Sept 8. Began Self-binder on Augt. 11th. and carried Harvest Home to-day!!! – the Best Hay, & Corn Harvest, I have seen in Oxon – for 22 years!!!

Sept 26. Paid a/c £5, breakfasted, & brot. Medicine away from Mr. Cheatle's.

Burford Fair – hired Beechey for Carter, House & garden, 13s. weekly, 50s. Michs. 10s. Hay & 10s. Harvest[5]

Sept 29. Mr. J.F. Maddox Sale, Shipton: tea'd with him!!

Witney Market Wheats 24s. to 26s. – Barleys 28s. to 34s. – Oats 18s. to 20s. – Consols 110¼ – The *best rain*, at night, for 12 weeks!!!!

This has been an exceedingly *Hot*, still, month!! . . .

Oct 6. Attended Mr. Geo: Bagnall, Westwell, Stock & Implement Sale – 500a. of his farm belongs Ch: Ch: College, Oxford; his son succeeds him in that occupation.

Oct 10. Shipton Auction – Lunch Mr. F.P. Matthews: Mrs. Price, there. Henry & John Franklin, Capps Lodge: both recently married.
Mr. Phillips, Oxford; bot. Mr. John Blake's freehold, in Cornmarket St. for £17,000.

Oct 12. Stratford on Avon Mop & Fair – 6 Oxen, & 14 pigs roasted in the Streets – Thousands attend, & help to eat them!!! . . .

Oct 14. The Duchess Marlborough confined of a 2nd son – most fortunate!!! . . .

Oct 28. Sir Herbert Kitchener, had a magnificent reception, on his arrival at Dover, and London – from Egypt.
Sir Herbert will really take the title of 'Lord Kitchener of Khartoum & Aspall', in the County of Suffolk!!!

Nov 17. Walked with Dawkins & Lord Moreton – shooting driven Partridges, shot 14 br. – Birds 8 Hares – 2 Phts. & Rabbit – bad shooting!!

Nov 19. Drove Fredc. Geo: to Mr. Sutton's, Wilcote: hunting – found in Holly Grove & killed – had good scents – killed 4 Foxes, last 3 days out!!! Wilson!!!

Dec 6. Major Scott, married Lillian Maud, the eldest Daughter of A. Brassey Esqr. M.P. Heythrop Park.
The G.W.R. ran a train from Paddington to Milford, 270 miles, in 5 hours 35 min. – only stopping 3 times at Newport, Landore, & Johnston!!!!

Dec 10. Drove, & met Hounds Eynsham Hall: Mr. Walter Ellis accd. John Charles, who broke his leg between Ankle & Knee, over a ditch & Hedge, at Hailey: a comminuted fracture of the tibia & fibula bones: John Chas. will have to lay stationary, for a week, or 10 days. . . .

Dec 26. Penny Post, to all our Colonies, commences to-day!!!
Wheat 27s. Barley 28s. to 34s. Oats 18s. Oranges 50 a shilling!! Consols $110\frac{1}{2}$.

Dec 31. 1898 has been a very *productive*, fine weather year; *all over* England!!

1899

Jan 26. Robt. Wm., drove me to shoot Mr. Sutton's covers at Wilcote, we shot 10 Pheasants, 4 Partridges, 4 Rabbits, & a Hare.

Witney Market, Red Wheats 28s. – Barley 28s. to 34s. – Oats 17s. to 19s.

Every prospect of a *record* year's Business in *General Trade*, Parliament – or rather Marquis Salisbury & Co., as *Popular* as ever – Opposition broken up – expect a useful year's home Legislation – London will be divided into about 30 Municipal Mayoralties & other *Gigantic* improvements done, for the working benefit, of the Colossal place. Only about 5 frosts during this month – a most mild winter – and the same all over Europe!!! . . .

March 4. A black Minorca laid an egg weighing 6 ounces, the largest I ever saw – also killed a Langchan Hen, weighing $8\frac{1}{4}$ lbs. – the heaviest I ever held up in hand!!!

March 20. Had 20 splendidly mild dry weather days, with $\frac{1}{4}$ inch snow – this morning, no wind, splendid clear sky, with bright sunshine – Work very forward in Farming – wheat down to 24s. & 25s. – Oats 17s.6d.

March 31. Walked to Priest Grove, Boynal, Nursery, admired brood phts. at Rangers Lodge – the Queen Beech – then down the *Grand Old* Beech Avenue to Cornbury House; the 150 Fallow Deer, looked handsome in the Park, so did the Gardens & Conservatories, also the range of Ponds.

March has been remarkably dry, mild, sunny, & still. . . .

April 10. Robt. Wm. fell from a Scotch Fir in Fairspear Chalkpit spinney, and broke his left arm & 3 left side ribs.

Dr. Kenny, Chipping Norton was called in. . . .

April 22. By 8.17 train from Ascott to Henley on Thames, 54 miles by rail, & only 16 and 22 – 38 by Cycle!!

April 23. Accd. Mrs. Abbott to Church – in p.m. Mr. Abbott walked me to top of Henley Park, saw Red & Fallow Deer & had a very fine view all round the County – saw Mr. Rigge's House – Evening at Mr. McBean's.

April 28. By 4.57 to Oxford – saw Dandy 5th. performed.

April 29. Saw Barnum & Bailey's wonderful procession, 40 Canadian Horses driven by one Man!! 16 Elephants – Prodigies – &c – Altogether a marvellous performance.

April 30. The 8th Duke Beaufort died in 76th year.

Wheat only 25s.6d. best red – Wool only 16s. – the two cheapest Articles of Produce in England – Consols 110¼.

A cold dry month. . . .

May 10. Mrs. Du Cros, buried at Finstock, ae 58 – from Cornbury Ho: her new Home!! . . .

Mr. Hiorns, Short Hampton dead ae 33.

Mrs. Padbury, Chipping Norton dead.

Mr. Geo. Cockrell from Crown Office called.

Henry Ferriman shorn the 4 rams: set in, extra warm.

May 16. By rail to Handborough, & attended County Agl. Show, in Blenheim Monument Park – Duke Marlboro', Earls Coventry & Jersey, Sir Walter Gilby, A. Brassey, &c.

Very satisfactory meeting – saw Lord Randolph's grave at Bladon.

May 17. Queen, laid foundation stone, of Victoria & Albert Museum, South Kensington.

With Wisdom, Goodness, Grace, she filled, for sixty years, the Throne
And whatso'er her People willed, that will, she made her own,
More long, more loved, she reigned than *all* the Kings of days gone by.
Sceptre may fade, and Empire fall, her name will never die –
Victoria, Victoria, long may she live & reign,
The Queen of our inviolate Isles, and Empress of the main!!

May 24. The Grand Queen-Empress Victoria's 80th birthday!! hale & sound!!

June 13. 45 years Marriage anniversary!!!

Ascot races commence to-day – Duke Marlborough & Mr. A. Brassey, drove their 4 in hands, on the Course.

June 20. Queen Victoria reigned 62 years – a record in History!!!

Had 5 dry, sunny, scorching weeks.

The Duke & Duchess York, spent 3 days at Oxford, with Dean Paget, attended the Encænia – Lord Kitchener of Khartoum, Mr. Cecil Rhodes, Lord Elgin, & 4 others; had D.C.L. conferred upon them – I saw them walk to the Sheldonian Theatre, & then to All Souls for Lunch – headed by the Duke, & Duchess of York, who was the *finest* woman amongst the large assembly.

Mr. Mrs. Cooper, Australia, taken Cornbury House, & shooting, for 2 years – what changes have occurred at Cornbury, the past 7 years!!!

June 30. June has mostly been a dry sunny month –

Wheat 25s. – Wool 16s. – Consols 107½. . . .

July 8. Using the *best* Rat-trap of my Life – a Patent wire one, only 3s. – sold by Fletcher & Phillipson, Dublin.

July 25. Binder began Winter Oats on 25th July; in 11 days cut & tied 118 acres of Corn – began to carry 11th Augt. – and finished 21st – 11 ricks. Splendid quality, & Condition, wheat weighs 66½ lbs per Bushel!!![1]

Aug 31. Wheat at Witney mkt. 23s. to 24s.6d. – Barley 27s. to 32s. – Oats 16s.

Queen starts for Scotland.

Ordered Mr. Dan Westell, to make my 'Will'.

Bought Robt. Willm. a Game Certificate, and a Gun Licence, for myself!!! after previously shooting 51 years!!!

There are 109 Deer Forests in Scotland average over 18,000 acres – a Stag is said to cost £30 – a Grouse a Guinea!!!

This is a record season for sport, amongst Deer & Grouse, Weather, *superb*!!!

Every prospect of War, or Trouble, amongst the Boers in South Africa – but Mr. Chamberlain will square the whole lot up, before 1900!!!

Cornbury Estate advertised for Sale – only £200,000 asked for it!!! – by Mr. Du Cros.

Sept 11. By 8.17 train to Oxford – reached Portsmouth 3 p.m. – via Reading, Basingstoke, & Winchester, alighted at Cosham Station for 4 Park Terrace, North End: found my Brother & Family well.[2]

Sept 12. Went over the Dockyard, saw the 'London' the *first of 6*, 15000 horse power War vessels, ready for launching.

Saw the 'Terrible' our fastest cruiser, preparing for starting to meet the 'Powerful' at the Cape –

Saw a very fine vessel building for the Japanese.

Sept 13. Sailed round the Harbour – inspected Nelson's 'Victory'.

Sept 14. By Packet to Ryde, was driven thro' the Osborne Estate, called & saw Prince Henry Battenberg's Tomb in Whippingham Church, went on to East Cowes, & Newport, & Carisbrooke Castle, & over the Downs to Ryde.

Sept 21. Returned to Langley, after spending 9 days at Portsmouth, which benefited my Health very much. . . .

Sept 25. Attended Burford Carter & Shepherd, hiring Fair, engaged Albert Hyde, House & Garden, 14s. weekly, & £1 for Haytime, £1 Harvest, & £1 extra at Michs.

John Charles & Wife spending 10 days at Eastbourne.

Burford Hiring Fair (Mop) 1895, taken by Henry Taunt. *(Oxfordshire County Libraries)*

Eynsham Diamond Jubilee celebrations, 1897. *(Oxfordshire County Libraries)*

Sept 30. Great preparations are being made by our Government, to *force* Mr. Kruger & South African Boers, & Orange Free State; to grant same laws & usages, to *all* alike, in those extensive Dominions!!!

Oct 12. War declared with the Transvaal & Orange Free State of South Africa – we shall win an Empire!!![3]

Wheat 30s. – Barley 27s. to 32s. – Oats 17s.

Oct 25. Attended Chipg. Norton, Board of Guardians.

Weather remains exceedingly dry & mild, we get no amount of Water, never knew the springs so low, had to haul *Drinking* Water, from Ford Wells Stone Trough!!!

Oct 27. Parliament prorogued after a 9 days Sessions: Voted 35,000 more soldiers, & £10,000,000 to carry on the war – more than 20,000 men are being sent from England alone, India & our Colonies, are all contributing – about 60,000 will be there next month.

Oct 30. Heythrop Hounds opening day – wet morn – dare not go – small meet – ran three Foxes, but did not kill.

Had 4 sharp engagements with the Boers at Glencoe, Elandslaagte, Reinfontein, & Ladysmith; Genl. Sir Redvers Buller has reached the Cape, expect when all are *there* & ready, the Boers will *wisely* throw up the Sponge: further fighting will be a reckless waste of Life!!! the Boers have fought *boldly*, & *bravely*, to their future great credit!!!

Nov 13. My 70th. Birthday – bot. 8 steers £10.10s. av. – returned to Langley and attended Shipton Auction in afternoon – Posted several Photos to my relatives, & Friends.

Nov 30. Much fighting has occurred around Mafeking, Kimberley, Colenso, Modder river, Belmont, Glaspan, Glencoe, Dundee, &c – the last three weeks: with every prospect of *much slaughter* this next month, over 17,000 persons are now there, ready for fighting in several directions; thro' Natal, into the Orange Free State, and the Transvaal: already over 3000 are slain and wounded!!! Sir Redvers Buller, is in Chief Command, along with Generals Lord Methuen, Clery, Gatacre, French, Hillyard &c – Hope much of it will be over by Xmas!!! Oh! dear, *no*.

Dec 12. . . . Our Cape Colony contains 276,000 sq. miles: the Transvaal, 129,000 sq. miles, & 25,000,000 acres.

We have been *much deceived*, by the *Number*, & *resources* of the Orange Free State, & Transvaal, *Boers*: it is a very *great* War – & will be, before they are subdued.

Dec 25. Xmas Day – Dined with my dau. Grace: Mrs. F.W.P. Matthews.

Dec 28. Wheat 25s.6d. Barley malting 27s. – Oats 17s. – at Witney.

The past year has been extra Dry, – & Sunny: a very light crop of Turnips, Autumn Pastures, & Clover aftermaths.

Dec 30. Quite a War craze – Volunteering to go to Africa, by thousands, this next year, ought to wind up War, for a long time: we shall have a *tremendous* inheritance, in Africa.

All kinds of Business, most Prosperous: except on the Land!!!

1900

Jan 6. Emily Ferriman died, ae 32 – a respected old Servant, and Labourer's wife. . . .

Jan 20. Our forces in S. Africa, checked at every attempt, to release our Soldiers, at Mafeking, Kimberley, & Ladysmith!!! and *all* by *Grazing Farmers*, riding a *Pony*, & using a *Mauser* Rifle.

Jan 25. Been a mild comparatively Dry month.
The War continues all the rage, Transvaal & Orange Free State, will be subdued – regardless of *Numbers*, or *Expence*, about 8000 dead, wounded, & missing, already!!! on our side.

Jan 31. Witney Mkt. Wheat 25s.6d. Barley 26s. – Oats 16s. – Bank rate 4½ p.c. Consols 100.
The Queen continues in the *Best* of health: so does the Duke Cambridge, who is 3 months older!!!
An immence trade doing, all over England.

Feb 16. More snow & rain the past *fortnight*, than we have had for years!!!
Kimberley relieved – Genl. French, peforming wonders!!!

Feb 24. Much fighting going on in Africa – losses up to *now*, officers killed 161, non-commissioned officers & men 1490, wounded officers 494, non-commissioned 5795 – missing and prisoners, officers 133, non-commd. & men 2669, *total 10,742.*

Feb 28. Consols 101 – Wheat 26s.6d. Barley 26s. to 28s. – malting: Oats 17s. to 19s. – seed. . . .
Our Generals carrying all before them in Africa – Kimberley & Ladysmith relieved on Tuesday and yesterday!!! to the great delight of *all* our *Friends*!!! 30,000 of our Colonials are assisting!!! what a grand example, and *assistance*!!!
The Queen renounced her visit abroad, owing to the War; going to spend a fortnight in Ireland, on 3rd April.

March 13. Bloemfontein entered; Boers all gone!! Lord Roberts clearing all the Orange Free State, as he passes along the Country for Pretoria.

March 17. Saint Patrick's Day, quite revived, much Shamrock worn!!! Met Hounds Cornbury: Mr. Ponsonby told me Mr. Percy Brassey was shot thro' the head & body on the day our Soldiers entered Kimberley; Mr. P.B. was in the 9th Lancers.

March 26. Albert Hyde, with a pair horses, Drilled 10a. Barley, never did the like before in my life!!! . . .

March 29. My 3 Sons, Grace, Isabel, & myself – all at Witney Mkt. Wheat 26s. – Barley 26s. – Oats 17s. & 18s. Consols 101.

March 31. Met Hounds at Blenheim, had Sherry, & Sandwiches – Duke in Africa, Duchess in London, their 2 Sons at Blenheim!!! Saw Mr. Angus the Steward, & Robertson, the Keeper.

The diary entries end at this point.

John Simpson Calvertt died on 2 May 1900 and was buried next to his wife in Leafield Church yard, about 10 yards from the south-west porch entrance.

Oxford Journal, Saturday May 5,1900, page 10:

'LANGLEY

AWFULLY SUDDEN DEATH OF MR. JOHN SIMPSON CALVERT.

The announcement of the very sudden and unexpected death on Wednesday morning of Mr. John Simpson Calvert, of Langley, will be received everywhere with deep sorrow. He was apparently in his usual vigorous health, and at 6.30 a.m. on Wednesday left his home for Witney. Shortly afterwards Maud Millins, domestic servant, found him lying in the passage quite dead. His face was downwards and a great deal of blood had flowed from the mouth. On the previous night, when he was seen by his son at 9.40, deceased appeared quite well. He was in his 71st year. Dr. O'Kelly was sent for from Chipping Norton, and the Coroner was informed.

For many years there was no better known figure in the local markets than the late Mr. Calvert, who for nearly half a century – 47 years, we believe – has been a farmer and breeder of eminence in our midst. He formerly farmed some 1,500 acres at Fairspeare and High Lodge, but with advancing age he felt the need of a quieter life, and surrendering Fairspeare to his son, he occupied the much smaller farm at Langley. Well informed, genial, kindly and courteous, he was held in very high esteem. He has long been the Guardian for Langley, and was present at the last meeting of the Chipping Norton Board of Guardians.'

Oxford Journal, Saturday May 12,1900, page 3:

'LANGLEY

THE LATE MR. JOHN SIMPSON CALVERTT.

An inquest was held on Thursday week at Langley House by Mr. Coroner Westell and a jury, of which the Rev. W. Collingwood Carter was foreman respecting the death of Mr. J.S. Calvertt, which was reported in last week's issue. The following evidence was given:–

Robert William Calvertt, son of the deceased, said his father was 70 years of age last birthday, and was a farmer living at Langley. Witness lived with his father. His father had suffered from a bad cough for about two years, otherwise he was in fair health. The last three or four days his breathing was, perhaps, a little worse than usual – he had a little difficulty in breathing. Witness saw his father on Tuesday night before he went to bed about 9.40, and heard him coughing for about five minutes after he had got upstairs. With that exception he heard nothing more of him during the night and he appeared to be in fairly good spirits when he went to bed. The next morning, at a quarter to seven, witness was called by Charles Ferriman, the groom, who came to his bedroom and said his father had been taken ill and was on the stairs, and asked him to put on his clothes quickly and come down. Witness at once dressed and went down, and found his father lying on the steps just outside the kitchen. His left hand was underneath him and his right hand was covered with blood. A large amount of blood had issued from his mouth. He was quite dead.

Maud Millins, domestic servant at Langley House, said she last saw her master alive on Tuesday night in the dining room, when he appeared quite well. She did not hear anything of him during the night. When witness came down at half-past six on Wednesday morning she saw a lot of blood on the floor near the scullery, and going through the kitchen she saw her master lying on the steps, dead. There was blood where Mr. Calvertt was lying. Witness called Miss Calvertt, who sent her for Charles Ferriman, the groom.

Dr. O'Kelly of Chipping Norton, said he attended the deceased for bronchitis last year. He had examined the body and found nothing remarkable upon it. The cause of death was hæmorrhage from the lungs.

A verdict in accordance with the medical evidence was given.

A correspondent writes: Mr. Calvertt was descended from an old established family, and was in his 71st year, having been born in 1829 at Witherne, in Lincolnshire. Losing his father, who was a surgeon, at an early age, Mr. Calvertt had some intention of following him in the medical profession, as he had acquired quite a knowledge of surgery, but being fond of a country life he went to learn farming for two years on the North Wolds, near Caistor, after which he occupied about 600 acres of land near Louth and Alford. He was considered a practical and model farmer, and greatly respected. He held many public offices, and was often appointed trustee. He had been a member of the Royal Agricultural Society for 43 years. Mr. Calvertt was considered one of the best game shots in Lincolnshire; he was at the large Tathwell Hall shoot, when the late Lord Henry Bentinck died suddenly from over-exertion and exhaustion. He was a bold rider, and a staunch preserver of foxes, his big wood being always a certain 'find'. It may here be mentioned that the South Wold hounds, often meeting at his house, ran through his wood nineteen times during one season. Wishing for a larger occupation and owing partly to his land being subject to floods Mr. Calvertt left Lincolnshire in 1876, and came to Fairspear Farm, of 1,200 acres, on the recommendation of the previous tenant, who had also been a native of Lincolnshire. Here he resided until 1895, when he resigned it to his son, and took the adjoining farm of Langley. It was a great undertaking for a man to leave his native county, where he was widely known and highly respected, to emigrate, as it were, 130 miles into a totally different district and class of land, and to make a new circle of friends. It shows what energy and determination and enterprise Mr. Calvertt possessed. In politics Mr. Calvertt was a sound Conservative, he took much interest in the erection of the Beaconsfield Hall at Shipton. He also wrote a daily record of his business, sport and pleasures, and his diaries, which extend over 50 years, are wonderfully interesting and show what a marvellously energetic man the writer was. Shooting all over the County, and hunting regularly, he yet found time to attend to his multifarious business duties, usually rising at 5 a.m. His death, though sudden and pathetic, was perhaps a merciful one, as he had an aversion to a bed of sickness. His courteous and kindly presence will be missed by many.'

Notes

1875–6

1. Lord Willoughby de Broke's steward at Holbeach.
2. Office of the Crown Commissioners of Woods, Forests and Land Revenues.
3. His two eldest sons: John Charles, born 28 June 1860; Robert William, born 6 March 1863. They were attending Christ's College, Finchley.
4. The Fairspear tenancy comprised Fairspear Farm [including Kingstanding Farm] 630a. 1r. 26p., and High Lodge Farm 559a. 2r. 30p., a total of 1190a. 0r. 16p. Calvertt and his family lived at Fairspear House.
5. The date on which Calvertt was due to take over the tenancy proper. The turnips and clover were to be consumed by livestock on the land, where their dung would later be ploughed in. Calvertt had no animals of his own on the farm during the long, slow process of moving from Lincolnshire, so he sold the fodder crops in this way in order to maintain the fertility of his land.
6. The total value of sheep bought was £1549.10s. Luck, or 'luck money' was an old custom whereby the seller gave the buyer a small extra sum in cash at the time of sale – a form of discount. The luck money would have brought the total down to £1547.10s., so Calvertt appears to have benefited by another £1.
7. Alice, born 27 May 1855; Elizabeth Anne, born 2 July 1857; Grace, born 12 August 1868; Frederick George, born 23 December 1873.
8. Charles Haddon Spurgeon was a Baptist Minister whose ministry was based upon the Metropolitan Tabernacle in Newington Causeway, London. He seems to have possessed enormous charisma and attracted huge crowds to his meetings.
9. Haymaking and harvest work was usually performed on a piecework basis and was let by farmers to groups of labourers or individuals at a flat rate per acre.
10. Boards of Guardians of the Poor were established by the 1834 Poor Law Amendment Act, which compulsorily amalgamated parishes into 'unions' to provide for the poor. The Guardians were the local instrument for Poor Law administration. They were required to have a property qualification and were elected by voters who also needed a local property qualification. Poor relief was financed by means of a Poor Rate, fixed by the Guardians, and levied on all local ratepayers. Fairspear Farm lay in Chipping Norton union and the union workhouse was situated in the town of Chipping Norton. Meetings of the Board of Guardians seem always to have taken place there.
11. Nearly every village had its one traditional holiday of the year, which extended over two or three days and incorporated several different events; dinners for the men, tea-parties for the women and children, parades, dancing and, sometimes, fireworks.
12. Leafield 5 years Club was a village benefit and sick club.
13. North Oxfordshire was a stronghold of the National Agricultural Labourers' Union, formed in 1872 by Joseph Arch and others. The general aim of the Union was to raise wages and improve conditions and, to this end, the labourers employed by Hambidge came out on strike on 21 April 1873 after their claim for higher wages had been rejected. Hambidge succeeded in

obtaining strike-breaking labour from the nearby parish of Ramsden, but on 12 May the womenfolk of the striking labourers tried to persuade the strike-breakers to stop work. Some of the women were carrying sticks, but this hardly justified the action of the Ramsden men, who enlisted police protection. As a result, the seventeen women were charged at Chipping Norton Petty Sessions with a breach of the 1871 Criminal Law Amendment Act. Sixteen of them were found guilty and sentenced: seven of them to ten days' hard labour, and nine to seven days' hard labour. Two of the women had babies at the breast and the children went to prison with their mothers. The cause of the women was quickly espoused by the national press, public opinion was outraged, and the 'Chipping Norton Affair' achieved a degree of notoriety. The women's fines were paid by public subscription and they were released from prison with a good deal of publicity, some of it directed against Hambidge.

14. When a steam ploughing company did contract work for farmers, the farmer often had to provide the coal and cart water for the engine. The Oxfordshire Steam Ploughing Company had its headquarters at Middle Cowley, Oxford. It was founded in 1868 and for a decade, concentrated solely on steam ploughing, but during the 1880's the business was extended to include steam rollers. The firm also manufactured steam ploughs and steam rollers.
15. A paternal relative.
16. These are further instances of piecework contracts. They illustrate the extent to which the contracts were linked to the condition of the crop.
17. The evangelical nature of Spurgeon's preaching seems to have been particularly attractive to the people of the area around Leafield.
18. The Council referred to was that of the Royal Agricultural Society of England.
19. Bliss's tweed mills still stand at Chipping Norton.
20. The Churchill baronetcy was created in 1815 for the younger son of the fourth Duke of Marlborough. Francis George Spencer, second baron Churchill, was born in 1802 and married Lady Jane Conyngham in 1849 – this is the Lady Churchill referred to by Calvertt.
21. It was common practice for farmers to pasture animals on corn stubbles. The valuable manure thus deposited was then ploughed in with the stubbles.
22. The men who worked with animals were the most essential of all farm workers to the farmer and their services were guaranteed for the whole year by means of hiring agreements. These agreements were frequently written, stating the precise terms of employment, just as Calvertt has stated them here, and were signed by both parties. It is interesting to note that a proportion of the wages of hired workers was paid in kind, and that a further proportion took the form of bonuses. Calvertt's examples are fairly typical of these agreements.
23. The Statute or Hiring Fair, or 'Mop'as it was commonly known in Gloucestershire and Oxfordshire, was the occasion on which shepherds, cowmen, grooms, carters, ploughmen, skilled labourers, milkmaids, dairymaids and housemaids displayed themselves in the market place for hire by prospective employers. By the nineteenth century these fairs had become a traditional holiday for farm workers, regardless of whether or not they were seeking employment.
24. Giddy: a dialect term applied to the effects of 'daft lamb disease' – a condition which is genetic in origin and where part of the lamb's brain [the cerebellum] is

congenitally atrophied.

25. Shoddy, the refuse woollen rags of shoddy cloth manufacture, was used as a fertiliser in some midland and northern counties of England – see James Caird, *English Agriculture in 1850–51*, 2nd edition, 1968, p 211. Mr. Aldworth may well have been using the waste product of the Witney blanket factories in a similar way.
26. Pedestrian races were extremely popular in the nineteenth century.
27. Calvertt presumably means 'tilth'.
28. The brothers Brassey, mentioned frequently by Calvertt, were the three sons of the railway magnate, Thomas Brassey [1805–70]. Sir Thomas Brassey [b. 1836] lived at Normanhurst, Battle, Sussex; Henry Arthur Brassey [b. 1840] lived at Preston Hall, Maidstone; Albert Brassey [b. 1844] lived at Heythrop Hall, Chipping Norton. The 'four young Brasseys' were the children of the latter.
29. Leafield School was a voluntary school and, as such, it was not eligible for aid from the rates. But the more influential Leafield ratepayers were obviously united in their wish to avoid the establishment of a school board under the terms of the 1870 Education Act. To this end they had raised a voluntary rate in order to provide a capital sum with which to remedy any deficiencies in buildings, equipment, etc., which might be reported by Her Majesty's Inspector of Schools.
30. Henry Chaplin was a substantial landowner who had been a neighbour of Calvertt's in Lincolnshire.
31. J.B. Lawes began to manufacture superphosphates in his factory at Deptford in 1843, basing his method on the research of Justus von Liebig, and thereafter began to market the artificial fertiliser commercially.
32. Burning and scalding were alternative methods of removing the bristles from pigs after slaughter.
33. Charles John Calvertt was his brother, born 23 November 1831.
34. The Sanger brothers staged many equestrian spectaculars in the 1860's, 70's and 80's, first at the Agricultural Hall, Islington and then at Astley's Amphitheatre. Miss Lavinia Sanger was the daughter of John Sanger.
35. Maskelyne and Cook's Seance was staged at the Egyptian Hall. In 1879 the attractions included: '. . . the sensational and marvellous DARK SEANCE, including Mr. Cook's astounding aerial flight in the cabinet from the stage to the dome in the centre of the hall', plus 'illusory sketches', and other entertainments of a similar kind. *The Times*, 13 February 1879.

1877

1. Calvertt hunted three times a week on average, usually with the Heythrop Hunt, but occasionally with the Vale of the White Horse or the Old Berkshire Hounds, for most of his life at Leafield.
2. It seems that none of the birds and ground game shot at Fairspear were wasted. Those not destined for domestic use or given as presents were sold to this Oxford butcher.
3. The Waywarden or Surveyor of Highways, was responsible for the maintenance and repair of local roads. The 1835 Highway Act empowered the levy of a highway rate, provided for the unification of parishes into a highway district [often coextensive with the Poor Law union] and allowed the payment of a district surveyor.

4. House guests from Lincolnshire.
5. The Russo-Turkish War had broken out earlier in 1877 when Russia had attacked Turkey, ostensibly to defend the Balkan Christians, who had suffered savage reprisals from the Turks following revolts against Turkish rule in 1875–6.

1878

1. In other words, the tenants received a rebate of ten per cent.
2. Calvertt was referring to his ordinary labourers, the men who performed the general work of the farm unconnected with the care of animals. These men were usually hired by the day and paid by the week and when there was no work for them, as in this situation, they were simply sent home or told not to report to work, and therefore received no wages.
3. The Congress of Berlin provided a forum for the European powers to meet, discuss and revise the terms of the treaty previously negotiated between Russia and Turkey.

1879

1. A relative of Jane Calvertt's, staying at Fairspear.
2. That is, had the loss advertised by the town crier.
3. The Zulu War had broken out on 12 January 1879. The news referred to by Calvertt was the battle of Isandhlwana which took place on 22 January, although news of the defeat did not reach England until 11 February.
4. This is another reference to Isandhlwana. The actual numbers of the dead were: 50 officers and 776 men of the 24th Regiment of South Wales Borderers and 800 men of various black contingents.
5. The wedding was that of Prince Arthur, Duke of Connaught, the seventh of Queen Victoria's nine children, to Princess Louise of Prussia.
6. That is to say, his landlord [the Crown] was giving him a rebate of £200 on the rent he had paid. Most, if not all landowners were eventually forced to help their tenants in a similar manner through the depression in corn prices and weather-induced disruption of the livestock breeding and rearing cycle. In the early years of the depression this usually took the form of a partial repayment of rent or an abatement of existing rentals, but by the 1890's most rents had been substantially reduced on a permanent basis.
7. This is a reference to the action at Myer's Drift which took place on 12 March.
8. This appears to be a reference to the action at Hlobane on 28 March.
9. Lady Churchill was Lady in Waiting to Queen Victoria.
10. This visit was part of a campaign undertaken by Joseph Arch between January and May 1879 in order to prevent a split within the National Agricultural Labourers Union [N.A.L.U.], as well as securing his own position as leader of that union. The campaign was largely successful but could not prevent a breakaway movement led by the secretary of the Oxford district of the N.A.L.U., Thomas Bayliss, and others. This movement was short-lived and unsuccessful. Pamela Horn, *Joseph Arch, The Farm Workers' Leader [1826–1919]*, 1971, pp 139–149.
11. The rent audit was the annual dinner given by landowners to their tenant farmers. It usually followed the last rent payment of the year.
12. That is, no remission or abatement of rent had been made to the tenants. Although the practice of making 'returns' was fairly general amongst larger

landowners in the early years of the depression, smaller landowners were less able or less willing to make such gestures to their tenants.

13. Prince Louis Napoleon, the Prince Imperial of France, was the only son of the late Emperor Napoleon III and Empress Eugenie, at that time resident in England. The Prince had trained at Woolwich and was keen to see some active service. He had gone out to Zululand with the reinforcements sent after Isandhlwana. Disraeli had not wished him to go, but could do little to stop him in the face of the active connivance of Queen Victoria and Empress Eugenie with the Duke of Cambridge which ensured that the Prince got his own way. He was killed on 1 June in a minor skirmish and the news caused a further slump in government prestige.
14. The lites, or lights, are still marked on the 1:50 000 Ordnance Survey map in the part of Wychwood Forest adjacent to Cornbury Park. Tower Light and Witney Light are named.
15. Calvertt is referring to a meeting of the Royal Agricultural Society.
16. Officials of the Crown Commissioners of Woods and Forests.
17. This is another example of harvest work performed on a contract basis by a group of men.
18. The battle of Ulundi took place on 3 July.
19. That is, grass keep; unsuitable weather conditions in the spring and summer had this undesirable result in the autumn, making the winter prospects for stock breeders and rearers rather grim.
20. This is a further example of the terms upon which the senior farm workers, the men who worked with animals, were hired. This example shows that farmers differentiated between married and single employees, often in the allocation of a tied cottage to a married man. The single men had to find their own accommodation, or arrangements would be made for them to board with married workers. Calvertt's carters were paid a 'harvest extra' bonus of 50s., in addition to their weekly wages, to cover the extra hours worked over the harvest period. The 20s. paid for 'Ale' was a money commutation of payment in kind with beer or cider, a practice increasingly frowned upon during the Victorian era.
21. Ewes afflicted with the 'rot', or liver fluke.

1880

1. This meeting was probably convened in order to discuss the forthcoming general election.
2. The 1880 general election resulted in a landslide victory for the Liberals, despite the split over Home Rule for Ireland. The Duke of Marlborough was consequently recalled from his post as Viceroy of Ireland, which he had held under Disraeli's Conservative administration since December 1876.
3. Under the Succession to the Crown Act of 1707 no M.P. could accept any important ministerial office without first vacating his seat and being re-elected. These by-elections were usually only a formality, but occasionally they resulted in government defeats. Since Harcourt continued as Home Secretary, a safe seat was obviously found for him at very short notice. The Act was not repealed until 1926.
4. Charles Bradlaugh was the Radical Liberal M.P. for Northampton. As an avowed atheist he claimed the right to affirm rather than swear the Oath of Allegiance in the prescribed form. Gladstone, whilst detesting Bradlaugh's

opinions, supported his right to affirm the Oath and take his seat as an M.P. But the House of Commons rejected the motion and Bradlaugh was expelled from the House. This happened repeatedly over the next few years: each time Bradlaugh was expelled he was re-elected by his Northampton constituents. The situation was not properly resolved until 1888, when an Affirmation Bill passed successfully through a Conservative House of Commons.

5. The Ground Game Act of 1881 authorised tenant farmers to destroy rabbits and hares on their farms without seeking their landlords' permission.
6. The reapers referred to here were, of course, reaping machines. The 5s. per acre quoted was the rate paid to the harvest labourers.
7. These were the daily wage rates for ordinary agricultural labourers.

1881

1. The Irish Campaign for Home Rule was being waged on two fronts at this time: in Ireland, by rent strikes and terrorism aimed at landlords; and in Parliament, against the Liberal government by the resources of debate and obstruction. Action on both fronts was orchestrated by Charles Stewart Parnell.
2. Liberal hesitation in granting independence to the Transvaal lead to the first Boer War. The action at Majuba Hill resulted in the defeat by the Boers of an inadequate British force sent from Natal. The outcome of the struggle was the granting of self-government to the Transvaal under the suzerainty of the British Crown.
3. When passed, this second Irish Land Act gave Irish tenants the right to fixity of tenure, fair rents [to be assessed by special courts] and free sale. But far from being an end to the Irish problem, it was simply another milestone, for Parnell calculated that if pressure on the government was maintained then still more concessions could be wrung from it.
4. President Garfield eventually died of his wounds on 19 September.
5. Calvertt's rent was £1,800 per year, so this was extraordinarily generous – few other landlords would be willing or able to make such a gesture, no matter how deep the agricultural recession had become.
6. Mr. C.A. Waller had obviously resigned the tenancy of the adjacent Crown holding.
7. The trees had been badly damaged in the gale of 14 October.

1882

1. Lord Cowper and Forster had resigned over the release of Parnell and two other Irish M.P.'s from jail as a result of the so-called 'Kilmainham Treaty'. Forster's replacement as chief secretary was Lord Frederick Cavendish, a younger brother of Lord Hartington married to a niece of Mrs. Gladstone. Burke was the under-secretary, a permanent official, and it was Burke who was the target of the attack in Phoenix Park – Cavendish suffered only because he was in Burke's company. The assassins belonged to the 'Invincibles', a small and obscure murder club. The Phoenix Park murders ended any hope Gladstone may have entertained of peace in Ireland.
2. Arabi Pasha, an Egyptian army officer, had led a revolt of his people against Anglo-French 'interference' in Egyptian economic and political affairs. When Arabi's followers began to fortify Alexandria and murdered some of the European inhabitants, the French withdrew and left the British ships assem-

bled off Alexandria to take action.

3. Ordinary labourers received none of the bonuses and payments in kind offered to the men who worked with animals. Calvertt demonstrates this quite clearly here: his labourers [who were in any case paid less than the hired men] had to pay rent for their farm cottages, while carters were given rent-free housing [tied cottages] in addition to other payments in cash or kind and bonuses.
4. Farmers generally paid Income Tax under Schedule B, which assessed their profits upon the amount of half their rental in England and Wales and one-third in Scotland and Ireland. In the event of not having realised this theoretical profit, farmers could attend the District Commissioners and obtain relief by the reduction of the charge to the amount of profit actually made – see: John Noble, *The Queen's Taxes*, 1870, pp 148–196. But as Calvertt had mentioned an assessment of £1320, clearly he had other sources of income which had been taken into account.
5. The closure resolution was moved by Gladstone on 3 February, after the House had sat for forty-one continuous hours, as a result of obstruction by Irish M.P.'s, while debating the Coercion Bill.

1883

1. By this Calvertt meant that it was always better to hold a tenancy on a large estate than on a small one. The very large landowners could afford to be a great deal more sympathetic and generous towards their tenants during this prolonged period of agricultural depression that the small owners with more restricted incomes and small reserves of cash. But even amongst the landed plutocrats, the Crown was an extraordinarily generous landlord by the standards of the day.
2. Ashton Court.
3. The addition of the self-binder to the reaping machine by the American firm of Walter A. Wood in 1876 represented the last stage of the development of the reaping machine in the nineteenth century. In the erratic harvest weather conditions described by Calvertt, the speed at which the reaper-binder worked substantially reduced the time needed by a farmer to get his crops cut, bound, stacked, carried and ricked, and enabled him to secure his harvest in the short dry periods in between the deluges.
4. Calvertt's previous farm was at Tothill Manor.

1884

1. She was visiting the Duke of Hesse.
2. It seems that Elizabeth Anne had become engaged to marry Mr. Abbott of Bombay earlier in the year, but the only indication of this was Calvertt's reference to the receipt of a letter from Mr. Abbott on 30 April.
3. The London Corn Market was held at Mark Lane.
4. Water supply was a constant problem in limestone areas.

1885

1. After Arabi Pasha's revolt, Egypt was left with a British military force in the country as a guarantee of order while Egyptian finance and administration were overhauled by British expertise. The Sudan was a dependency of Egypt but, when a leader who proclaimed himself the Mahdi, or Messiah, stirred the

people into revolt, Gladstone declined further involvement in Egyptian affairs and ordered the evacuation of the marooned Egyptian garrisons in the Sudan. General Charles Gordon was selected for this task but, far from retreating, he attempted to defy the Mahdi. By May 1884 Gordon had been cut off in Khartoum by the Mahdi's forces. A relief force set out from Cairo in September, but did not arrive at Khartoum until 28 January 1885. Khartoum, starved into exhaustion, had fallen two days earlier and Gordon had been killed during the six hours of massacre that had followed. The event did little to improve Gladstone's already battered public image and incurred Queen Victoria's open displeasure.

2. Emily Calvertt was John Simpson Calvertt's niece, the daughter of his brother Charles, who lived at Huntingdon. Only five bridesmaids were named – the sixth should have been Miss Nellie Bullock.
3. His rent was previously £1,800 per year.
4. While British attention was concentrated on the Sudan, the Russian government had decided that the time was ripe for an extension of its own dominion into Afghanistan. Gladstone reacted with vigour on this occasion, securing a vote of £11,000,000 from Parliament and putting the army and navy on the alert. The Sudan, meanwhile, was abandoned to the Mahdi. Diplomacy proved to be the right course of action: war with Russia was avoided, and Afghan and British pride remained unruffled.
5. William Warn owned the Barton Steam Brewery, Church Street, Tetbury.
6. Salisbury's Conservative government lasted for seven months and was strictly a 'caretaker' administration. The Conservatives had no majority in the House of Commons and carried on the business of government only until an election could be held.
7. The effect of the third Parliamentary Reform Act [1884] was to equate the county with the borough franchise, giving the vote in the counties to all householders and adding about 2 million voters to the existing 3 million. This explains the pre-election activity in places such as Leafield which had not previously merited the attention of the two party machines.
8. Calvertt still owned property at Louth and Huntingdon and was, therefore, eligible to vote at both places.
9. The Primrose Tory League was created in Autumn 1883 by Lord Randolph Churchill. Its purpose was ostensibly to provide voluntary political effort at and between elections, but it was really part of Churchill's attempt to create a personal following and to break the Conservative leadership in his own favour. In 1884 the League was officially recognised as part of the Conservative establishment, thus robbing it of much of its potency as a vehicle for individual political ambition. In this particular instance, campaigning by the Primrose Leaguers was obviously regarded by the Liberals as having been over-zealous.

1886

1. The 86 Irish Nationalist M.P.'s held the balance of power in the House of Commons after the general election. They knew that Salisbury's Ministry did not favour Home Rule for Ireland and therefore used their vote to defeat the Conservatives as soon as the new House met. Gladstone, at heart, favoured Home Rule although he would have preferred to solve the problems of Ireland through inter-party agreement. But when he became Prime Minister for the third time in 1886 he was dependent on Irish votes to stay in office and so

virtually pledged to give Ireland Home Rule, although many Liberals were opposed to the idea.

2. Meetings and marches of the unemployed were organised by the Social Democratic Federation [S.D.F.], a socialist organisation founded in 1881. As a result of this Trafalgar Square meeting, four notable socialists were prosecuted but were acquitted in April after a four day trial. Similar disturbances occurred in Manchester and elsewhere. Perhaps the events of January in Leafield were not entirely unconnected. Rick-firing [10 February below] was a traditional form of rural protest.
3. Lord Randolph Churchill was the first politician to see the significance of Ulster in the Home Rule struggle and he deliberately exploited the Orange faction in order to stir up resistance in Ireland and at home. In Belfast he made some inflammatory speeches, coining the phrase "Ulster will fight and Ulster will be right" in one of them. Having stirred up a hornet's nest in Ulster, he retreated back to England.
4. Ruston and Hornsby Ltd., Sheaf Ironworks, Lincoln: manufacturers of portable and traction engines.
5. This effectively reduced his rent for the year from Michaelmas 1885 to Michaelmas 1886 to £785, or less than half his 1875 rent of £1800 per year.
6. Jane Calvertt's mother.
7. Calvertt must have made this entry a few days later, since he has the wrong day and the wrong figures. The Home Rule Bill was defeated on its second reading on 8 June by 343 votes to 313. Some 93 Liberals voted against the Bill. Gladstone resigned at once and Parliament was dissolved. The Home Rule issue not only brought down Gladstone's Ministry in this instance, it also split the Liberals and, apart from one brief interval, consigned them to opposition for the next twenty years.
8. Calvertt had recorded the news that Elizabeth Anne was expecting a child on 22 October 1885, but had not recorded his birth.

1887

1. Positive discrimination in favour of British-grown corn had disappeared with the Corn Laws in 1846. One of the distinctive features of the agricultural depression of the late nineteenth century was the comparative inability of British corn to compete in price with cheap imports. Not surprisingly, this revived the demand amongst farmers and landowners in corn-growing areas for the return of protective measures.
2. Calvertt seems to have become confused, William Cody and Buffalo Bill were one and the same person.
3. Hayward Simpson of Icomb Place, Icomb.
4. This was the result of a shortage of grass keep and fodder crops induced by the summer drought. Most graziers and rearers were trying simultaneously to sell off the stock they could not feed over the coming winter and, consequently, prices fell.
5. Oxfordshire had substantial deposits of clay suitable for the manufacture of bricks, tiles and pottery. Leafield had been the site of several potteries since medieval times, but the 'China Shop' referred to by Calvertt was probably the Crown Pottery. This pottery was established in the late 1850's or 1860's, probably by the Franklin family, and there they manufactured a variety of earthenware pottery: bread and milk pans, jugs, honey-pots, tobacco jars, and

ornamental flower pots, until the late 1890's. During the 1890's the lease appears to have passed to Mr. Groves, the builder and contractor of Milton-under-Wychwood, while the Franklin family carried on at the pottery as employees. The making of pottery ceased around 1898 and from then until the sale of the site in 1921, only bricks, tiles and drainage pipes were manufactured. Oxfordshire Museums Service, Publication no. 13, *The Clay Industries of Oxfordshire: Oxfordshire Potters*, 1980, pp 21–24.

6. Mr. Samuda appears to have been a landowner who kept abreast of agricultural innovation. Silage making was introduced into Britain from the Continent via America in the early 1880's. Initially the reception was favourable and by the late 1880's about 1000 silos had been constructed in Britain. But inadequate research, insufficient technical knowledge, and failure to standardise silo design soon alienated many farmers, who preferred to adhere to haymaking as a source of winter fodder, despite the unreliable nature of the British climate.

   The reference to calf rearing may indicate that Mr. Samuda was experimenting with an intensive calf rearing system in a purpose-built rearing house complete with stalls, rationed feeding, and proper manure conservation methods. Such systems often met with disapproval, since their cost was high in terms of capital and labour and the more elaborate the system, the higher the cost.
7. This activity may have been the result of the passing of the Allotments Act of 1887.

1888

1. It seems that Leafield voluntary school had met with an adverse Inspector's report in 1887 and in the September of that year the Committee had passed a resolution requesting the master to resign. At the meeting held on 2nd May the committee members had proposed and seconded a motion of dissatisfaction with the master, but the chairman, the Reverend Lee, refused to put it to the meeting claiming that the committee was not properly constituted. The committee therefore got up and left. Leafield was a voluntary school, and if it failed then a Board School would have to be established. School boards were generally unpopular with ratepayers on the grounds of extra cost and other disadvantages, real or imagined.
2. It is difficult to believe that this is the 'Kaiser Bill' of the First World War.
3. For some reason none of Jane Calvertt's nurses seem to have stayed long at Fairspear.
4. An old friend of the Prices who had died some months before, after moving to Torquay.
5. A house guest, a friend of Grace.
6. This would appear to have settled the dispute which had arisen earlier in the year.
7. The most important reform brought in by Salisbury's Ministry of 1886–92 was the Local Government Act of 1888. This legislation set up County Councils – elected bodies which took over the duties previously performed by nominated Justices of the Peace. It was a measure of landowners' power in rural areas that it had taken country people over 50 years to achieve the same control over local government that borough residents had enjoyed since the Municipal Corporations Reform Act of 1835.

8. In other words, Potters Hill Farm was 'in hand' – being run by the Crown estate until a tenant could be found.

1889

1. Unfortunately, Mr. F. Matthews failed in his attempt to become a County Councillor, losing to S. Groves by 4 votes.
2. Presumably Calvertt means shallots.
3. Lowther Henry Barker was a close friend of Calvertt's from Lincolnshire.

1890

1. This was a fund set up by Queen Anne in 1704 to receive and use the annates and tenths previously confiscated by Henry VIII. The revenue from the fund was usually used to supplement the incomes of the poorer clergy.
2. The dockers' victory in the London Dock Strike of 1889 and the trade boom of 1889–90 combined to give trade unionism a stimulus. One of the most distinctive features of this increased trade union activity was the organisation of unskilled labour. This 'new unionism', which manifested itself in many places in England and Wales, had two important features: it organised men by industries, and it preached political as well as industrial action. Its leaders – men like Ben Tillett, Tom Mann and John Burns, were conscious socialists, and it was their efforts which did much to pave the way for the formation of the Labour Party in 1900.
3. Frederick had left Louth Grammar School in the previous year after a bout of illness. All the Calvertt children seem to have changed their schools fairly frequently.
4. Lincolnshire.
5. Parnell had been cited as co-respondent in a divorce case brought by Captain O'Shea against his wife, Kitty, who had been Parnell's mistress for several years. The revelation was a severe shock to Parnell's Catholic following, as well as the nonconformists who constituted an important element in the Liberal party. Parnell stubbornly refused to resign his leadership of the Irish Nationalist Party and it simply broke up around him.
6. A regular house guest.

1891

1. It seems as though Calvertt rather frowned on this match – Mr. Price was 68 and Miss Pearson 24.
2. See 1880 footnote 4 above.
3. A new battleship, launched by Queen Victoria a few days earlier.
4. Calvertt's Lincolnshire cousins.
5. Calvertt had not recorded the birth of a second son to the Abbotts, the first having died in 1886 at the age of eight and a half months.
6. Calvertt was probably referring to the post of assistant overseer of the poor. The overseers had once been important in Vestry administration but the 1834 Poor Law Amendment Act had transferred their levying and supervisory duties to the guardians of the poor. Their residual function was merely the assessment and collection of poor rates.
7. It was Isabel's turn to be despatched to boarding school.
8. He was A.E. Abbott's son by a previous marriage and therefore Elizabeth Anne's stepson.

9. A feud seems to have developed between Calvertt and William Mace after the 'sheep stealing' incident of February 1890.

1892

1. Albert Victor, Duke of Clarence, was the eldest son of the Prince of Wales.
2. Viscount Dillon, Ditchley Park.
3. Calvertt was a member of the Royal Agricultural Society.
4. These were his resident farm workers – those who lived in the farm cottages.
5. Calvertt apears to be a little confused here – the lady was most probably married, but it is difficult to determine which was her married name.
6. The Conservatives and Liberal Unionists together won more seats than the Liberals in this election, but the way in which Calvertt arranged the figures shows that it was the Liberals, in alliance with the Irish party, who formed the government. It was Gladstone's intention to introduce a second Home Rule Bill, and he took office as Prime Minister for the fourth time with this aim.
7. Grace and her husband, Mr. and Mrs. F.W.P. Matthews.
8. This, presumably, was the reason for the change of farms. He seems to have blamed High Lodge Farm for the greater part of his financial losses and therefore decided not to renew the lease when it expired. Langley Farm obviously seemed to have greater potential.

1893

1. The Bill passed its three readings in the House of Commons, but was eventually defeated by the large Conservative majority in the House of Lords.
2. By this time Lord Randolph Churchill was a sick man, his brain disease well advanced. He had been asked by Balfour to return to the Conservative front bench in the Commons at the end of 1892, but he was of no use to the party leadership and his Commons performances were pathetic. Lord Rosebery later said, "he died by inches in public". P.R. James, *Lord Randolph Churchill.*
3. Mr. Adams had died at Hailey Manor on 6 March, aged 77.
4. As with Elizabeth Anne, Calvertt had omitted to record the birth of Alice's second child.

1894

1. Gladstone was prevailed upon to resign as Prime Minister by his colleagues following the defeat of his second Irish Home Rule Bill in the House of Lords.
2. Bernard was left with his paternal aunt when his parents returned to India.
3. Frederick George had taken a job with the Coventry cycle manufacturers, Rudge & Co.
4. This was the first time that John Charles' engagement to Mary Simpson had been mentioned by Calvertt.
5. These were: her son, the Prince of Wales [Edward VII]; her grandson, the Duke of York [George V]; and the new baby, later named Edward – the future Edward VIII, the uncrowned king.
6. The Tithe Commutation Act of 1836 commuted the tithes due to the rector of the parish into an annual rent. This situation prevailed in many places until the Tithe Act of 1925 abolished the rent charge, and the Tithe Act of 1936 extinguished them altogether.
7. Calvertt is describing the wall above the porch at Langley House. The Eleanor of Guienne he referred to should presumably be Eleanor of Aquitaine. The

initials can still be seen over the porch of the house. They are not the original ones, however, but are later replacements, possibly dating from the renovation of Langley House when the Crown estate was carved out of Wychwood Forest in 1857–8.

8. The Sino-Japanese War of 1894–5 was the culmination of a struggle for supremacy between the two nations in Korea. The war was a total disaster for the Chinese, who suffered military and naval defeats, followed by economic and territorial humiliation in the Treaty of Shimonoseki in 1895.
9. Urban and rural district councils and parish councils, based upon direct popular election, were created by the Local Government Act of 1894. The civil duties of the parish vestries were transferred to the new parish councils.
10. Lord Randolph Churchill, sick though he was, had embarked upon a world tour earlier in 1894. His health collapsed when he reached Madras and he was brought home at the end of the year.

1895

1. The 'Crathie' of Aberdeen, a steamer of 475 tons, had collided with the 'Elbe', a North German Lloyd passenger steamer of 4510 tons en route from Bremen to New York, in a heavy sea and near gale force wind at 5.30 a.m. on 30 January, 47 miles west of the Haaks lightship in the North Sea.
2. Bertha Dix had clearly been hired by Calvertt as a domestic servant and the £14 was her yearly wage.
3. The Baltic Canal, begun in 1887, linked Brunsbüttel on the North Sea coast with Holtenau on the Baltic coast.
4. Gladstone had resigned as Prime Minister and retired from the Liberal front bench on 2 March 1894, after the defeat of his second Irish Home Rule Bill in the House of Lords. The Liberals had great difficulty in replacing him and his successor, Lord Rosebery, was neither his choice nor the party's, but Queen Victoria's. Rosebery proved to be ill-fitted to hold the Liberals together and his Ministry was short-lived.
5. I Zingari was a club of an exclusive nature formed in 1845, devoted in the main to cricket and the theatre.
6. The Agricultural Holdings Act of 1875 gave legal force to the customary practice of compensating the outgoing tenant for the unexhausted value of improvements made to the land. These included temporary improvements in the form of acts of cultivation and manuring, the consumption of purchased feeding stuffs on the farm and the application of purchased manures. In certain cases the improvements could also include work of a more long-term nature, such as draining, liming and chalking.
7. Consols: government stock.
8. The 1895 general election resulted in a decisive victory for the Conservatives and Liberal Unionists.
9. The marriage was far from being the success that Calvertt hoped for. It was largely a marriage of convenience for both families: the ninth Duke of Marlborough was badly in need of money to restore Blenheim to its former glory, while the wealthy Vanderbilts wished to gain entrée into English high society. After 15 years of separation, the marriage ended in divorce in 1921.
10. Bully, or bullace: a wild plum, a larger variety than the sloe.
11. Calvertt's summary of the state of British foreign relations at the end of 1895 reflects not only the rising tide of British imperialism at this point, but also the

estrangement of Britain from most of the major world powers. President Grover Cleveland of the United States had officiously intervened in a dispute between Britain and Venezuela over the boundary of British Guiana, a gesture motivated as much by the vote-winning appeal of 'twisting the lion's tail' in the forthcoming presidential elections as from a desire to see the dispute settled in America's best interests. Britain was not really involved in the Cuban problem, but the British government had decided to support the American view of the conflict between the United States and Spain over Cuba.

Africa loomed large in British foreign affairs. The reference to Italy and the Abyssinians concerned the beginning of moves towards British reconquest of the Sudan between 1896 and 1898. A Franco-British confrontation in Africa had been threatening since 1894: the main areas of conflict were in West Africa, where French expansionist policy had led to competition with Britain for control of the Niger, and in Egypt. Ashanti lay right across the approaches from the British Gold Coast to the exterior and the activities of Ashanti warriors had made progress in the form of trade and 'civilising missions' virtually impossible. The British government responded to this situation by sending out a military expedition to Ashanti under the command of Colonel Sir Francis Scott late in 1895.

1896

1. Relations between the British and the Boers in South Africa had worsened after the badly mistimed attempted coup in the Transvaal, masterminded by Cecil Rhodes and led by L.S. Jameson, in the last days of 1895. The Boers entertained well-founded suspicions that the Foreign Secretary, Joseph Chamberlain, an ardent imperialist, was an accomplice in the Jameson raid.
2. Master and matron of the workhouse.
3. The Agricultural Rates Act of 1896 remitted one-half of the farmer's rates. It was the first major example of the 'de-rating' principle adopted by parliament.
4. Mr. Robert Abraham, Minster Lovell, had died on 16 October, aged 76.
5. Calvertt seems to be referring to a reduction in the freight charges fixed by railway companies. Many farmers, landowners and traders had been arguing for some years that railway freight rates favoured the bulk transport of imported produce from ports to urban areas and discriminated positively against the domestic producer. The railway companies, on the other hand, insisted that consigments of farm produce from rural railway stations were irregular, badly-packed, and too small.

1897

1. Miss North was Robert William's fiancée.
2. Mr. Beauchamp was an official of the Crown Office of Woods, Forests and Land Revenues.
3. Calvertt was obviously unable to remember the correct name at the time of writing – it was John Chaundy.
4. Tithes were divided into Great or Rectorial tithes and Small or Vicarial tithes in parishes where the rector and the incumbent, or vicar, were different people. After the dissolution of the monasteries about one-third of the rectors were laymen, who appointed vicars to take their place. Under the Tithe Commutation Act of 1836 tithes could be commuted to a rent charge based on the prevailing price of corn. What Calvertt intended to show here was the extent

of the depreciation of the rent charge since it was first fixed.

5. The firm of estate agents who had originally offered the Cornbury estate for sale in 1895 had taken Lord Churchill to court in an attempt to claim the full commission that they felt was due to them from the sale. During the course of the court action it transpired that the estate had been sold in 1896 to Mr. Du Cros and an associate, Mr. Hooley, for £150,000: a price considerably in excess of the reserve put upon it at auction earlier in the year. It was also claimed that the introduction of the buyer to the seller had been effected by a third party, and not by the estate agents or any person connected with them. Judgement was given in Lord Churchill's favour.

1898

1. The strike was called by the Amalgamated Society of Engineers and involved several issues, the most prominent amongst them being the demand for an eight hour day. The employers organised themselves against the strike into the Engineering Employers' Federation, which began on the Clyde but under its leader, Colonel Dyer, became a powerful national organisation. The long and bitter struggle began in July 1897 and dragged on until 15 January 1898, when the men surrendered. The defeat of the most powerful union in Britain had profound effects upon the world of labour, stimulating some of its organisers into a preference for political rather than trade union action, and others into strengthening the union movement by building larger units through the fusion and federation of existing unions.
2. In 1896 Salisbury's government decided to gain security for Egypt and control of the upper waters of the Nile through conquest of the Sudan. The campaign, which lasted until 1898, was directed by Sir Herbert Kitchener. Sirdar was the title given to the British commander-in-chief of the Egyptian army.
3. Mr. Hooley was an associate of Harvey Du Cros, and co-owner of the Cornbury estate. Their joint purchase of the property would seem to indicate that their business interests were closely linked.
4. This entry was obviously made by Calvertt a few days later.
5. The 50s. was a bonus paid at Michaelmas the following year – in other words, it was a bonus paid upon completion of the hiring contract. The '10s. Hay' and '10s. Harvest' were bonuses paid at the end of the two harvest periods, in recognition of the longer hours worked at these seasonal work peaks.

1899

1. Most of this entry was added at a considerably later date.
2. Calvertt had not mentioned his brother's move from Huntingdon.
3. The British public confidently expected the second Boer War to be over in a few weeks.

Glossary

Beast: a generic term for cattle.
Charlock: Field Mustard.
Crew: a yard with sheds for cattle.
Cuckoo lamb: a lamb born late in the spring.
Drag: the drag harrow was a strong cultivating harrow with curved tines. The harrow was mainly a surface-working implement, used to obtain a fine tilth and to cover seeds after sowing or drilling. But harrows did perform other work and a considerable number of different types, such as the drag harrow, were constructed.
Drape ewes/cows: old animals no longer fit for breeding, or barren animals.
Eddish: grass or clover grown again [after cutting] for a second crop.
Fagging: or bagging. A form of hand reaping, performed with a heavy fagging or bagging hook, rather than a reaping hook or a sickle. The corn was cut near the base of the stalk with a slashing action.
Fallow: to leave the land bare, ploughing or cultivating it at intervals to keep it free from weeds.
Gelt ewes: barren ewes.
Half-bred: cross-bred.
Keep/keeping: grazing for livestock.
Landrail: the Corn-crake, a migratory species of bird appearing in Britain in late April and departing in mid- or late September.
Lease: sportsman's term for a brace and a half – three creatures of any kind.
Mop: Fair usually held at or near Michaelmas [29 September] or, occasionally, Lady Day [25 March]. It provided an opportunity for farm and domestic servants to find new employers and for farmers to find new servants.
Mowing [manual]: to cut grass or corn crops with a scythe. During the last half of the nineteenth century the scythe was used increasingly to cut corn crops, as it was faster than using sickles, reaping or fagging hooks.
Plash: to plash, pleach or lay a hedge is to make, dress or renew the hedge by cutting the stems partly through, bending them down, and interlacing the stems, branches and twigs so as to form a close, low fence which would in time grow in height.
Rot: liver fluke, an acute or chronic disease affecting sheep of all ages, caused by the ingestion of a parasite which uses wet or muddy land during part of its life cycle.
Sainfoin: A leguminous fodder crop.
Sheep: there was great regional variation in the terms used to describe sheep at different ages, but the following definitions can be taken as broad indicators:–

Lamb: birth to autumn [sex usually denoted by a prefix e.g. ram- or ewe-, he- or she-; although when ram- and he- were used simultaneously he-lambs were usually castrated lambs].

Teg/Hog/Hogget: between autumn and first shearing [sex often denoted by a prefix, as above].

Shearling/Shearhog: between first and second shearing. With shearling, the sex was often denoted by the use of the relevant word e.g. shearling ram, ewe, or wether.
Theave: female sheep after second shearing.
Ewe: fully-grown female.
Ram/Tup: fully-grown male.
Wether: fully-grown castrated male.

Single: to thin out a growing crop.
Skerry: or scuffler, an implement for weeding turnips.
Small seeds: clover and grass seed.
Twitch: couch-grass.
Twitching: cleaning the land of couch-grass.
Vetch: leguminous fodder plant.
Yean: to lamb.

Metric money equivalents

6d.	=	2½p	10s.	=	50p
1s.	=	5p	15s.	=	75p
5s.	=	25p	19s.	=	95p

Pre-metric measurements

1. Length [Imperial]
   1 rod, pole or perch = 16½ feet or 5½ yards
   40 poles = 1 furlong = 220 yards
   8 furlongs = 1 mile = 1760 yards
2. Area [Imperial]
   30¼ square yards = 1 square pole = 272¼ square feet
   40 square poles = 1 rood = 1210 square yards
   4 roods = 1 acre = 4840 square yards
3. Weight [Imperial]
   14 pounds [lb.] = 1 stone [st.]
   28 pounds = 1 quarter [qr.]
   4 quarters = 1 hundredweight [cwt.]
   20 hundredweight = 1 ton
   N.B. Butcher meat in London was sold by the Smithfield stone of 8 lbs., but in some counties of England the stone contained 12, 14, or even 16 lbs. Imperial.
4. Wool weight
   14 lbs. = 1 stone
   2 stone = 1 tod = 28 lbs.
5. Dry solid capacity
   1 peck = 20 lbs.
   4 pecks = 1 bushel = 80 lbs.
   8 bushels = 1 quarter = 640 lbs.

Index